SpringerBriefs in Applied Sciences and Technology

SpringerBriefs present concise summaries of cutting-edge research and practical applications across a wide spectrum of fields. Featuring compact volumes of 50 to 125 pages, the series covers a range of content from professional to academic.

Typical publications can be:

- A timely report of state-of-the art methods
- An introduction to or a manual for the application of mathematical or computer techniques
- A bridge between new research results, as published in journal articles
- A snapshot of a hot or emerging topic
- An in-depth case study
- A presentation of core concepts that students must understand in order to make independent contributions

SpringerBriefs are characterized by fast, global electronic dissemination, standard publishing contracts, standardized manuscript preparation and formatting guidelines, and expedited production schedules.

On the one hand, **SpringerBriefs in Applied Sciences and Technology** are devoted to the publication of fundamentals and applications within the different classical engineering disciplines as well as in interdisciplinary fields that recently emerged between these areas. On the other hand, as the boundary separating fundamental research and applied technology is more and more dissolving, this series is particularly open to trans-disciplinary topics between fundamental science and engineering.

Indexed by EI-Compendex, SCOPUS and Springerlink.

Anand Raj • Praveen Nagarajan
Shashikala A. P. • Sudha Das

Rubcrete and Its Applications in Structures Subjected to Impact Loading

Anand Raj
Department of Civil Engineering
Chulalongkorn University
Bangkok, Thailand

Praveen Nagarajan
Department of Civil Engineering
National Institute of Technology Calicut
Kozhikode, Kerala, India

Shashikala A. P.
Department of Civil Engineering
National Institute of Technology Calicut
Kozhikode, Kerala, India

Sudha Das
Department of Civil Engineering
National Institute of Technology Calicut
Kozhikode, Kerala, India

ISSN 2191-530X ISSN 2191-5318 (electronic)
SpringerBriefs in Applied Sciences and Technology
ISBN 978-981-95-6314-2 ISBN 978-981-95-6315-9 (eBook)
https://doi.org/10.1007/978-981-95-6315-9

This Springer imprint is published by the registered company Springer Nature Singapore Pte Ltd.
The registered company address is: 152 Beach Road, #21-01/04 Gateway East, Singapore 189721, Singapore

About This Book

This book focuses on rubcrete and its fibre-reinforced variants, their principles, characteristics, development, testing, and practical applications in structural elements subjected to impact loading. Rubcrete is an innovative and sustainable material that partially replaces mineral aggregates with crumb rubber derived from waste tyres. When reinforced with fibres (steel or polypropylene fibres), it offers a balanced improvement in strength, ductility, and energy absorption capacity.

Rubcrete represents a promising alternative to conventional concrete by addressing two major challenges: enhancing impact resistance and promoting environmental sustainability. The reuse of waste rubber not only reduces landfill disposal but also contributes to resource conservation and a reduced carbon footprint in the construction sector. The material's inherent flexibility and toughness make it suitable for critical applications such as prestressed railway sleepers, crash barriers, fencing posts, and kerbs, where resilience to impact is essential.

The book systematically presents the development, mechanical performance, and durability behaviour of rubcrete and its fibre-reinforced variants, supported by experimental and numerical studies. It further introduces the concept of energy absorption to cost ratio, providing a practical basis for evaluating the efficiency of modified concretes in real-world conditions.

In summary, this book bridges the gap between sustainable material innovation and structural engineering practice. It serves as a valuable reference for researchers, practising engineers, and postgraduate students interested in advanced concretes, sustainable materials, and resilient infrastructure design.

Contents

Chapter 1
Rubcrete

1.1 Tyre Waste: A Growing Environmental Concern

The population of the world is on the rise. So is the need for transportation. The number of vehicles being registered generally rise with rise in population. As the most populous country, in India steadily rising population has also been accompanied by a significant increase in vehicle production, as evidenced by the data published in the Road Transport Yearbook (2017–2018 and 2018–2019) [1] (Fig. 1.1). This surge in vehicular growth inevitably leads to a corresponding escalation in tyre consumption and disposal, raising critical concerns regarding end-of-life tyre management.

At the end of their functional lifespan, worn-out tyres frequently find their way into landfills, unregulated dumping grounds, or unauthorised storage sites. This unchecked accumulation presents a multifaceted environmental challenge. Discarded tyres, in addition to being visual blights on the landscape, pose significant fire hazards. Tyre fires are notoriously difficult to extinguish and can burn for days, releasing thick plumes of toxic fumes that compromise air quality and endanger human health. Moreover, when exposed to the nature, these tyres often become receptacles for rainwater, inadvertently transforming into ideal breeding grounds for mosquitoes and other disease-carrying elements. The health risks associated with such stagnant water storage are particularly severe in densely populated regions. Beyond these immediate hazards, the slow degradation of rubber in open environments leads to the leaching of harmful chemicals and hydrocarbons into the surrounding soil and groundwater, contributing to long-term ecological damage. The vulcanised nature of tyre rubber, engineered for durability and resistance to decay, renders it virtually non-biodegradable. As a result, waste tyres persist in the environment for decades, earning them the designation of one of the most stubborn solid wastes in contemporary waste management systems.

A. Raj et al., *Rubcrete and Its Applications in Structures Subjected to Impact Loading*, SpringerBriefs in Applied Sciences and Technology,
https://doi.org/10.1007/978-981-95-6315-9_1

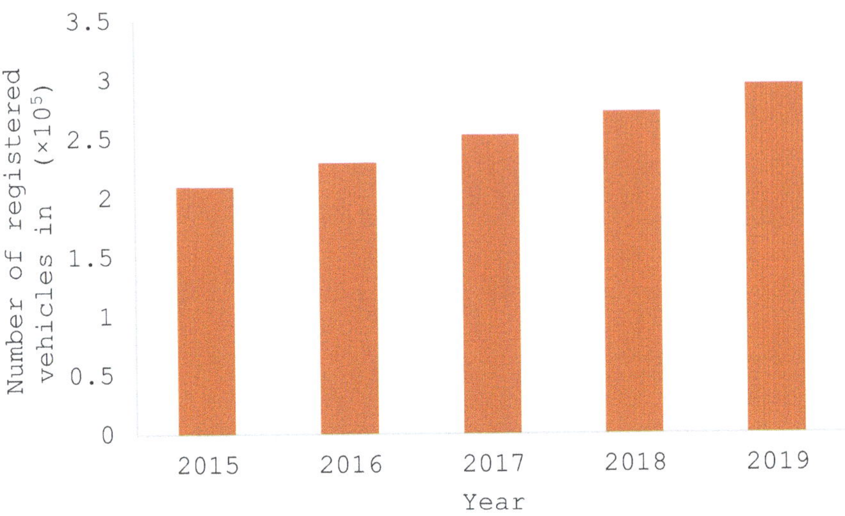

Fig. 1.1 Details of registered vehicles in India (2015–2019)

1.1.1 Existing Disposal Practices and the Need for Reuse

Conventional disposal methods, such as landfilling, incineration, and uncontrolled stockpiling, not only fail to address the core of the problem but often aggravate it. The fire risks associated with storage, coupled with the toxic emissions from incineration, highlight the urgency of finding more sustainable pathways. Recycling and reusing end-of-life tyres have emerged as an environmentally responsible alternative for the above problems. However, effective waste mitigation requires more than just recycling for its own sake. The transformed materials must be purposefully reintegrated into engineering applications where their properties can be meaningfully exploited.

Yet, to achieve true circularity and long-term sustainability, recycling must not be a passive act of diversion from landfills; it must lead to a purposeful reincarnation of waste materials. In this regard, the construction industry holds remarkable potential. Among its most resource-intensive components, concrete offers a fertile ground for innovation, particularly through the substitution of conventional mineral aggregates with recycled alternatives. This opens a gateway to harnessing end-of-life tyres, not merely as waste, but as a valuable engineering resource.

Fig. 1.2 Transformation of tyres into aggregates

1.2 Alternative Aggregates for Sustainable Waste Management

Natural mineral aggregates have long been indispensable in concrete production, serving as the bulk component in terms of both volume and structural function. However, with the escalating pace of urbanisation and infrastructure development, the availability of quality natural aggregates is becoming increasingly strained. The global demand for aggregates has reached a level where even manufactured sand and recycled concrete aggregates are now being routinely explored to be incorporated into mix designs to mitigate resource depletion.

Amidst this growing scarcity, discarded tyres present an innovative and ideal alternative for aggregates. The transformation process begins with the mechanical stripping of steel reinforcements embedded in the tyres, resulting in large torn fragments of rubber. These fragments, when cut into coarser pieces, are referred to as shredded rubber which are suitable for use as partial replacements for coarse aggregates. When further processed into finer particles, the material becomes crumb rubber, which can serve as a substitute for fine aggregates. This dual adaptability positions tyre-derived rubber aggregate as a credible alternative to conventional aggregates, bridging the gap between waste mitigation and sustainable material sourcing. More importantly, it establishes an elegant synergy: where one resource (waste rubber) is diverted from an environmental burden to meet the shortage of another (natural aggregates). In doing so, it marks a promising leap toward more sustainable construction practices. Figure 1.2 represents the transformation of tyre rubber into aggregates.

1.3 Rubcrete

Rubcrete refers to a form of concrete in which conventional mineral aggregates, both coarse and fine are replaced by rubber aggregates derived from discarded tyres. This innovative composite material draws upon two forms of processed rubber: shredded rubber, obtained from coarsely cut tyre chunks after the removal of steel reinforcements, and crumb rubber, a finer, powdered form produced through further processing.

In practice, shredded rubber is typically employed as a partial replacement for coarse aggregates, while crumb rubber substitutes a portion of the fine aggregates (sand). This substitution facilitates the diversion of significant volumes of tyre waste from landfills, aligning well with sustainability goals. However, the mechanical performance of rubcrete is notably influenced by the type and proportion of rubber used.

Experimental studies have consistently shown that the introduction of rubber particles leads to a reduction in strength characteristics, including compressive, tensile, and flexural strength [2–9]. Among the two types, the detrimental effect is more pronounced when shredded rubber is used to replace coarse aggregates. The irregular geometry, low stiffness, and weak bonding interface of large rubber chunks undermine load-bearing efficiency, rendering such mixes less suitable for structural applications.

In contrast, crumb rubber, with its finer gradation and improved dispersion within the cement matrix, demonstrates better compatibility with structural concrete. Although it too results in strength reduction, the effect is comparatively moderate and can be controlled through optimisation of mix design. As such, crumb rubber emerges as the more viable candidate for structural applications, offering a balanced compromise between sustainability and performance.

1.4 Structural Applications

Given the modest compressive strength of rubcrete, it may not be ideal for primary load-bearing members in high-rise construction. However, in components where impact resistance, ductility, and resilience are the primary design requirements, rubcrete emerges as a technically and environmentally compelling alternative. Some key application areas of rubcrete in Impact-Sensitive Infrastructural Elements discussed in this book include:

- **Prestressed Concrete Railway Sleepers**
  Railway sleepers serve as foundational elements that transmit and distribute the load from passing trains. They endure repeated mechanical shocks and must maintain track stability under extreme conditions. Incorporating rubcrete into prestressed concrete sleepers enhances their energy dissipation capacity, damping behaviour, and resistance to crack propagation. Moreover, the use of recycled rubber contributes to a more sustainable rail infrastructure by reducing dependency on natural aggregates and reducing tyre waste.
- **Concrete Fencing Posts**
  Fencing posts installed in rural, industrial, or highway settings often endure impact from livestock, vehicles, and flying debris. The addition of rubber particles into these posts improves their shock absorption capacity, enhances toughness, and prevents brittle failure, making them ideal for locations where reliability under sudden impact is critical.

- **Crash Barriers**
  Concrete crash barriers play a pivotal role in highway safety. Designed to redirect or contain vehicles during high-speed accidents, these structures must possess exceptional impact resistance and post-crack ductility. Rubcrete, with its high energy absorption and flexibility, can mitigate the severity of collisions and reduce barrier damage under repeated impacts, especially when subjected to high-speed or heavy-load vehicular strikes.
- **Concrete Kerbs**
  Functioning as both protective and aesthetic boundaries along roadways, kerbs often encounter impact from vehicle tyres, accidental collisions, and construction machinery. Rubcrete improves their ability to withstand repeated low-energy impacts without fracturing, thereby extending the operational life of kerbs and reducing maintenance demands.

The procedure for mix design of rubcrete and its applications are discussed in the following chapters.

References

1. Ministry of Road Transport and Highways Transport Research Wing, Road Transport Yearbook (2017–2018 & 2018–2019). New Delhi (2021)
2. A. Raj, P. Nagarajan, S. Aikot Pallikkara, Application of Fiber-reinforced rubcrete for crash barriers. J. Mater. Civ. Eng. **32**, 04020358 (2020). https://doi.org/10.1061/(ASCE)MT.1943-5533.0003454
3. A. Raj, P. Nagarajan, S. Aikot Pallikkara, Application of Fiber-reinforced rubcrete in fencing posts. Pract. Period. Struct. Des. Constr. **25**, 04020037 (2020). https://doi.org/10.1061/(asce)sc.1943-5576.0000512
4. A. Raj, P. Nagarajan, A.P. Shashikala, Behaviour of fibre-reinforced rubcrete beams subjected to impact loading. J. Inst. Eng. (India) A (2020). https://doi.org/10.1007/s40030-020-00470-4
5. A. Raj, P. Nagarajan, A.P. Shashikala, Failure prediction of impact behaviour of self-compacted rubcrete sleepers. Mater. Des. Process. Commun., 1–8 (2020). https://doi.org/10.1002/mdp2.174
6. A. Raj, P. Nagarajan, A.P. Shashikala, Failure prediction of impact behaviour of self-compacted rubcrete sleepers. Mater. Des. Process. Commun. **3**, e174 (2021). https://doi.org/10.1002/mdp2.174
7. A. Raj, P. Nagarajan, S.A. Pallikkara, Performance of fiber-reinforced rubcrete curbs subjected to impact loads. Pract. Period. Struct. Des. Constr. **27**(3) (2022). https://doi.org/10.1061/(ASCE)SC.1943-5576.0000707
8. A. Raj, C. Ngamkhanong, L. Prasittisopin, S. Kaewunruen, Nonlinear dynamic responses of ballasted railway tracks using concrete sleepers incorporated with reinforced fibres and pre-treated crumb rubber. Nonlinear Eng. **12**, 320 (2023). https://doi.org/10.1515/nleng-2022-0320
9. A. Raj, P. Nagarajan, S.A. Pallikkara, C. Ngamkhanong, Performance characterization of fiber-reinforced rubcrete sleepers. J. Struct. Des. Constr. Pract. **30** (2025). https://doi.org/10.1061/JSDCCC.SCENG-1743

Chapter 2
Composition of Rubcrete

2.1 Rubcrete

This chapter presents the development of rubcrete and its fibre-reinforced variants, aimed at improving the impact resistance of concrete structures. A systematic approach has been adopted to study the material composition, mix design procedure, and key performance characteristics of rubcrete. The chapter begins by describing the materials used in the preparation of concrete mixes.

Following the material overview, the chapter outlines the development of mix proportions for concrete of different strength. A key component of this study is the procedure adopted to find the optimum percentage of crumb rubber and fibre content. To determine the most effective and economical proportions, drop-weight impact tests were conducted. The performance of each mix was evaluated using a cost-to-benefit ratio, where the benefit is defined in terms of impact energy absorption capacity. Based on this analysis, optimum contents were identified for:

- Crumb rubber in rubcrete
- Steel fibres in steel fibre-reinforced concrete (SFRC) and rubcrete
- Polypropylene fibres in polypropylene fibre-reinforced concrete (PFRC) and rubcrete

The following engineering properties of the optimum mix were determined:

- Compressive strength
- Flexural strength
- Stress–strain relations and strain energy density
- Fracture energy

Subsequently, the durability performance of the optimised mixes was studied through:

A. Raj et al., *Rubcrete and Its Applications in Structures Subjected to Impact Loading*, SpringerBriefs in Applied Sciences and Technology,
https://doi.org/10.1007/978-981-95-6315-9_2

- Water absorption
- Sorptivity
- Resistance to acid, sulphate, and marine water exposure

Through this structured investigation, this chapter builds a comprehensive understanding of the development and performance of rubcrete and its fibre-reinforced forms, setting the foundation for their application in impact-prone structural elements.

2.2 Mix Design

Concrete is a composite material whose properties are governed by the type and quality of its constituents. Various materials were selected based on functional requirements such as strength, workability, impact resistance, and durability. This section provides a detailed account of all the materials used in the development of fibre-reinforced rubcrete mixes for concrete of minimum compressive strength of 20, 40, and 60 N/mm^2. The materials include cement, metakaolin, natural sand, crushed stone, crumb rubber, steel fibres, polypropylene fibres, potable water, superplasticisers, high-range water reducers, and polyvinyl alcohol. The sample calculations for the determination of mix proportions are given in Appendix.

2.2.1 Cementitious Materials (Binders)

Cementitious materials form the binding phase in concrete. They are responsible for initiating hydration reactions, providing strength, and holding the composite together. A combination of Portland cement and metakaolin was used for concretes of different strengths.

Ordinary Portland Cement (OPC) and Portland Pozzolana Cement (PPC)
Cement acts as the primary binding agent. When combined with water, it forms a paste that coats the aggregates and undergoes hydration, forming calcium-silicate--hydrate (C-S-H) the main phase responsible for strength in concrete. This hydration product creates a dense, interconnected matrix that contributes to compressive strength and overall durability.

The choice of cement was based on the required compressive strength of concrete:

PPC (IS:1489 Part 1: 1991) [1] was used in concrete of compressive strengths 20 and 40 N/mm^2. It offers better resistance to sulphate and chemical attack, making it suitable for general applications.

OPC 53 Grade (IS:12269: 2013) [2] was selected for concrete of strength 60 N/mm^2 due to its higher early strength, which is essential for prestressed concrete structural elements.

The physical properties of both cement types are presented in Table 2.1.

Table 2.1 Properties of cement and metakaolin

Property	OPC (53 grade)	PPC	Metakaolin
Standard code	IS: 12269–2013	IS: 1489 (part 1)–1991	–
Specific gravity	3.14	3.14	2.6
Initial setting time (min)	~30	~30	–
Final setting time (min)	~600	~600	–
Fineness (m^2/kg)	~320	~300	Very high
Bulk density (kg/m^3)	~1100	~1000	400–500
pH value	–	–	4.5–5.5
Pozzolanic reactivity	Low	Moderate	High
Strength of concrete	Concrete of strength 60 N/mm^2	Concrete of strength 20 and 40 N/mm^2	SCM in concrete of strength 60 N/mm^2

Metakaolin (MK)

To further enhance the performance of high-strength concrete, metakaolin was used as a supplementary cementitious material (SCM). Metakaolin is obtained by calcining refined kaolin clay at controlled temperatures, resulting in a highly reactive pozzolanic material. Properties of metakaolin are shown in Table 2.1.

Key benefits of metakaolin include:

- Strength enhancement through additional C-S-H (Calcium-Silicate-Hydrate) and C-A-S-H (Calcium-Alumino-Silicate-Hydrate) formation.
- Improved durability by refining the pore structure and reducing permeability.
- Mitigation of alkali–silica reaction (ASR) and reduction in efflorescence.
- Early age strength development and lower drying shrinkage.
- Sustainability, as it replaces a portion of cement and reduces embodied carbon.

2.2.2 *Fine Aggregates (Sand)*

Fine aggregates play a vital role in shaping both the fresh and hardened properties of concrete. Often underestimated, they contribute not just to workability but also to the strength, density, and long-term durability of the mix. Far from being inert fillers, fine aggregates serve multiple structural and functional purposes.

Enhancing Density and Reducing Voids One of the primary functions of fine aggregates is to fill the voids between coarse aggregate particles. This filler effect significantly enhances the packing density of the mix, reducing the internal voids within the concrete. A denser aggregate skeleton leads to fewer weak points, resulting in:

- Increase in compressive strength
- Reduced permeability, thereby enhancing durability
- Improving workability and cohesion

Fine aggregates contribute to the workability of fresh concrete by acting as a lubricant in the mix. They help in the uniform distribution of materials, reduce segregation and bleeding, and enhance mix cohesion. A well-graded fine aggregate comprising a good balance of particle sizes minimises water demand and ensures easier placement and finishing.

Reducing Cement and Water Demand By efficiently filling internal gaps, fine aggregates help reduce the quantity of cement paste required to bind the mix. This not only makes the mix more economical but also allows for:

- A lower water-to-cement ratio
- Higher strength and durability of the hardened concrete
- Controlling shrinkage and cracking

The inclusion of fine aggregates in the concrete matrix helps in restraining drying shrinkage, which is inherent to cement paste. Their even distribution aids in controlling micro-crack formation and reduces the risk of early-age shrinkage cracking, thereby improving the structural integrity.

Contributing to Strength and Surface Bonding Fine aggregates support strength development both directly and indirectly:

Directly, through mechanical interlock and bonding with the cement matrix.

Indirectly, by enabling lower water content and reducing porosity.

The surface texture and angularity of the fine aggregate particles influence this bond. While angular particles enhance mechanical interlocking, they can marginally reduce workability.

Supporting Aesthetics and Finish In applications requiring a smooth finish such as kerbs, fencing posts, and architectural concrete, the texture and colour of the fine aggregates influence the visual and tactile quality of the final surface.

Properties of Fine Aggregates Locally available manufactured sand was used as the fine aggregate for the development of the mixes. It had the following properties: specific gravity = 2.675, fineness modulus = 2.86, grading = confirming to zone II as per IS 383: 2016 [3]. The particle size distribution of the fine aggregate used is shown in Fig. 2.1.

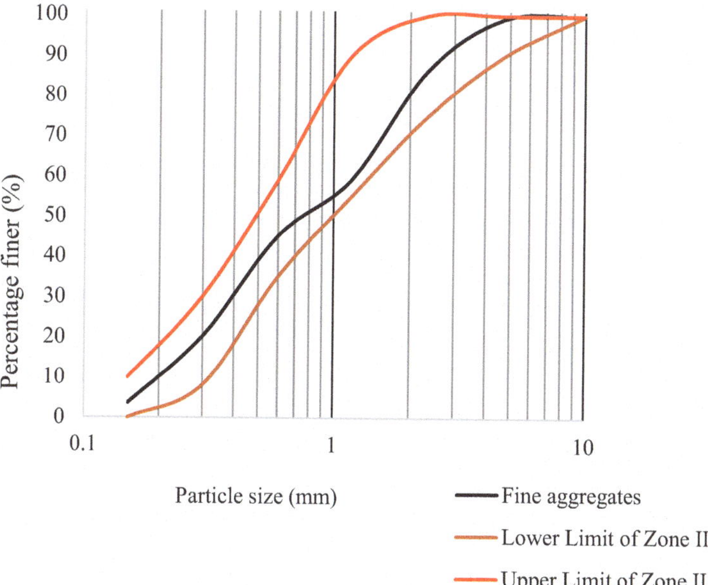

Fig. 2.1 Particle size distribution of fine aggregates

2.2.3 *Coarse Aggregates*

Coarse aggregates form the structural backbone of concrete. Constituting approximately 60–80% of the concrete volume, they significantly influence its mechanical behaviour, dimensional stability, and long-term durability. Apart from being fillers, coarse aggregates are vital contributors to the strength, workability, and economy of the mix.

Providing Bulk and Reducing Cost Coarse aggregates allow for a substantial reduction in the use of cement paste, which is the most expensive component of concrete. By occupying the majority of the mix volume, they contribute to:

- Economic efficiency in large-scale construction
- Volume stability, without compromising structural performance
- Enhancing compressive strength

Coarse aggregates serve as internal load-bearing elements. They form a rigid skeleton that transmits compressive stresses effectively through the hardened concrete matrix. Given that the aggregates themselves are typically stronger than the surrounding cement paste, failure usually initiates in the interfacial transition zone (ITZ) rather than in the aggregate particles.

The strength contribution depends on:

- The quality and hardness of the aggregate
- Its particle shape and texture

- The degree of bonding with the cement matrix
- Angular, rough-textured aggregates are known to improve mechanical interlocking and bond strength compared to rounded or smooth ones
- Dimensional stability and crack resistance

Coarse aggregates play a crucial role in restraining drying shrinkage and thermal movements. Their relatively stable dimensions prevent the paste from contracting excessively during the hydration and drying processes. This reduces internal tensile stresses and the likelihood of cracking, thereby enhancing the long-term integrity of the structure.

Improving Durability A well-graded aggregate blend, combined with a suitable binder system, results in a dense, impermeable concrete. This improves the concrete's resistance to:

- Water and chemical ingress (chlorides, sulphates)
- Freeze–thaw cycles
- Abrasion and erosion, especially in pavements, industrial floors, and marine applications
- The use of durable, non-reactive aggregates also reduces the risk of deterioration under aggressive environmental exposure

Influencing Workability and Mix Economy The maximum size and gradation of coarse aggregates directly affect the workability and economy of concrete:

- Larger aggregates reduce the total surface area to be coated by paste, allowing lower water and cement demands.
- However, excessively large aggregates may cause segregation, reduce finish ability, and interfere with reinforcement in heavily congested sections.
- A well-graded aggregate system with an appropriate mix of large and small particles enhances packing density and allows for smoother placement.

Properties of Coarse Aggregates Two sizes of crushed stone were used as coarse aggregates, selected based on the target strength of concrete: 12 mm nominal size for concrete of strengths 20and 40 N/mm^2 and 6 mm nominal size for concrete of strength 60 N/mm^2, to suit the closely spaced reinforcement layout in prestressed concrete sleepers. The coarse aggregates used had a specific gravity of 2.79. The particle size distribution of both types of coarse aggregates is shown in Fig. 2.2.

2.2.4 Water and Superplasticisers

Water is a fundamental component in concrete, essential for both hydration and workability. Potable water, conforming to IS 456: 2000 [4], was used for all mixing and curing purposes.

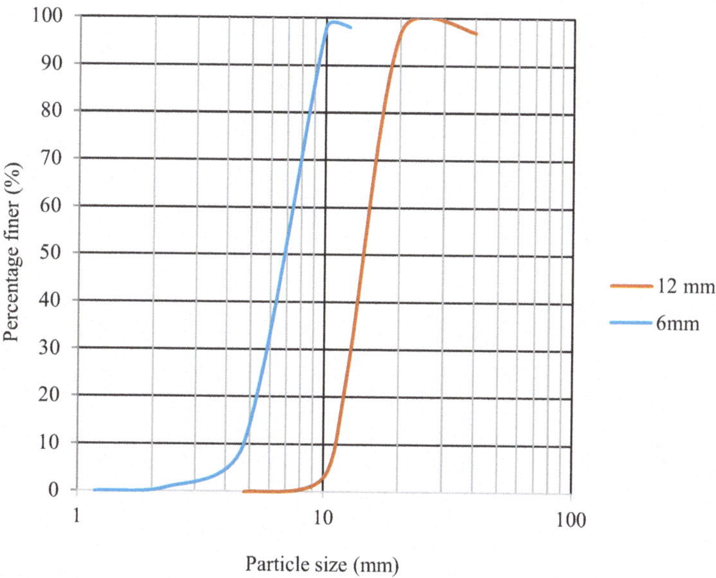

Fig. 2.2 Gradation curves for coarse aggregates (12 mm and 6 mm)

The water-to-cement ratio is a critical parameter influencing the strength, durability, and permeability of concrete. However, achieving the desired workability without increasing the water content requires the use of chemical admixtures most notably, superplasticisers and high-range water reducers (HRWRs). The superplasticisers and HRWRs used for the concrete of different strengths are described below.

Concrete of Strength 40 N/mm² To improve the workability of concrete of strength 40 N/mm², a sulphonated naphthalene polymer-based superplasticiser was used. This admixture improves fluidity and dispersion of cement particles, thereby reducing the water demand while maintaining the required slump. Table 2.2 presents the key properties of the sulphonated naphthalene-based superplasticiser [5].

Concrete of Strength 60 N/mm² For concrete of strength 60 N/mm², where both high-strength and self-compacting behaviour were required, a combination of HRWR and viscosity modifying agent (VMA) was adopted. The HRWR ensured sufficient workability at low water-cement ratios, while the VMA helped stabilise the mix, preventing segregation and bleeding. Table 2.3 summarises the properties of the HRWR and VMA.

Mix Ratio of Ordinary Concrete At this point, it is important to understand that cement, fine aggregates, coarse aggregates, and a combination of water and superplasticisers form the basic constituents of concrete. Which means an ideal combination of these ingredients would help to develop a concrete with desired strength and workability. Let us delve into the process of development of a concrete of minimum

Table 2.2 Properties of sulphonated naphthalene-based superplasticiser

Property	Value
Appearance	Dark brown liquid
Specific gravity	~1.20 at 25 °C
pH range	7.0 ± 1.0
Chloride content	<0.2%
Dosage range	0.5–2.0% by weight of cement
Function	Improves workability, reduces water demand

Table 2.3 Properties of HRWR and viscosity modifying agent

Property	HRWR	VMA
Appearance	Light brown liquid	Transparent viscous liquid
Specific gravity	~1.08	~1.02
pH value	6.0–8.0	6.0–8.0
Chloride content	<0.2%	Nil
Function	Water reduction, workability	Segregation control, stability
Dosage range	0.3–1.5% of binder content	0.1–0.3% of binder content

Table 2.4 Mix proportions of ordinary concrete

Mix ID	Cem	Met	FA	CA	Wat	HR	VM	Rub	SF	PF
M60R0	1	0.0930	1.519	2.700	0.367	0.009	0.005	0	0	0

Cem- cement; *Met*- Metakaolin; *FA*- fine aggregates; *CA*- coarse aggregates; *Wat*- water; *HR*- high-range water reducer; *VM*- viscosity modifying agent; Rub-Crumb rubber; *SF*- steel fibres; *PF*- polypropylene fibres

strength 60 N/mm^2 which would be suitable for prestressed concrete sleepers. The mix design of this mix is discussed in the Appendix of this book. Table 2.4 presents the combinations of ingredients that would provide a mix with a compressive strength of 60 N/mm^2.

In Table 2.4, the Mix ID M60R0 denotes that the intended compressive strength of the mix is 60 N/mm^2, and 0% of the volume of fine aggregates is replaced with equal volume of crumb rubber. Now let us find out what happens when we replace manufactured sand with crumb rubber so as to enhance the energy absorption capacity.

2.2.5 *Crumb Rubber*

Crumb rubber (Fig. 2.3) is a recycled material derived from discarded automobile tyres [5]. It can be used as a partial replacement for fine aggregates, aimed at enhancing the impact resistance and energy absorption capacity of concrete. Its application

Fig. 2.3 Crumb rubber

not only contributes to sustainable construction but also addresses the growing problem of tyre waste disposal.

Crumb rubber serves a dual purpose in concrete:

Material substitution: By replacing a portion of the fine aggregates, it helps reduce the demand for natural sand, thereby preserving mineral resources.

Energy absorption: Rubber particles are deformable and elastic. Their inclusion improves the toughness and ductility of concrete, making it more resilient under impact and dynamic loads.

To maintain structural performance, care must be taken to match the gradation of rubber particles with that of natural sand. This ensures uniform distribution, minimises segregation, and facilitates adequate bonding within the concrete matrix.

The crumb rubber used was obtained from a local tyre retreading facility. The physical properties were as follows: specific gravity = 0.65, texture = rough and angular, and colour = dark grey to black.

A sieving process was employed to ensure that the particle size distribution (Fig. 2.4) of the crumb rubber closely matched that of the fine aggregates it replaced. This similarity in gradation was critical to maintain workability and performance consistency.

Pre-treatment of Crumb Rubber Crumb rubber particles were pretreated by immersing them in a 2% polyvinyl alcohol (PVA) solution, prepared using 20% of the total mixing water. The rubber was soaked in the PVA solution for 30 min to improve bonding with the cement matrix and reduce water absorption. This was done to ensure that the surface of the powdered crumb rubber particles is free from the dirt and debris that may have accompanied them.

Mix Ratio of Rubcrete Now that we have discussed the fundamental properties of crumb rubber, let us focus on replacing 5%, 10%, 15% volume of sand with equal volume of crumb rubber. Table 2.5 shows the mix proportions of rubcrete.

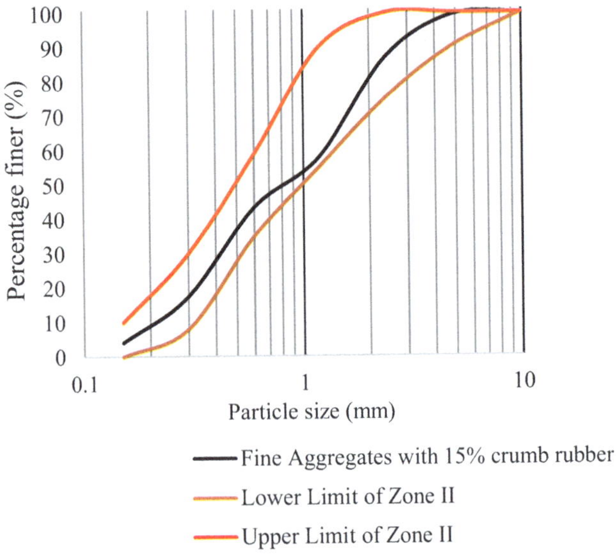

Fig. 2.4 Particle size distribution curve of fine aggregate added with 15% crumb rubber

Table 2.5 Mix proportions of rubcrete of concrete of strength 60 N/mm^2

Mix ID	Cem	Met	FA	CA	Wat	HR	VM	Rub	SF	PF
M60R5	1	0.093	1.442	2.701	0.367	0.009	0.005	0.018	0.000	0.000
M60R10	1	0.093	1.367	2.701	0.367	0.009	0.005	0.036	0.000	0.000
M60R15	1	0.093	1.290	2.701	0.367	0.009	0.005	0.054	0.000	0.000
M60R20	1	0.093	1.215	2.701	0.367	0.009	0.005	0.075	0.000	0.000

Optimum Percentage of Crumb Rubber The replacement of crumb rubber beyond 15% has resulted in significant decrease in compressive strength (Fig. 2.5). Hence the limit of replacement of manufactured sand with crumb rubber was fixed at 15%. Let us now understand the reason for the reduction in compressive strength of rubcrete when crumb rubber is added. The addition of crumb rubber imparts voids in the concrete material matrix. These voids weaken the inter-material bonding, which breaks when subjected to gradual compressive force. This results in a reduction in compressive strength with the addition of crumb rubber.

The reduction in compressive strength with the addition of crumb rubber is evident from Fig. 2.5. The compressive strength reduced by 3%, 6%, 10%, and 25% when the crumb rubber percentage was varied by 5%, 10%, 15%, and 20%. Now the next question is 'how much of crumb rubber is ideal for rubcrete for optimum energy absorption capacity?'

To ensure both cost-efficiency and mechanical performance, the optimal proportion of crumb rubber in the rubcrete was determined using an energy absorbed to

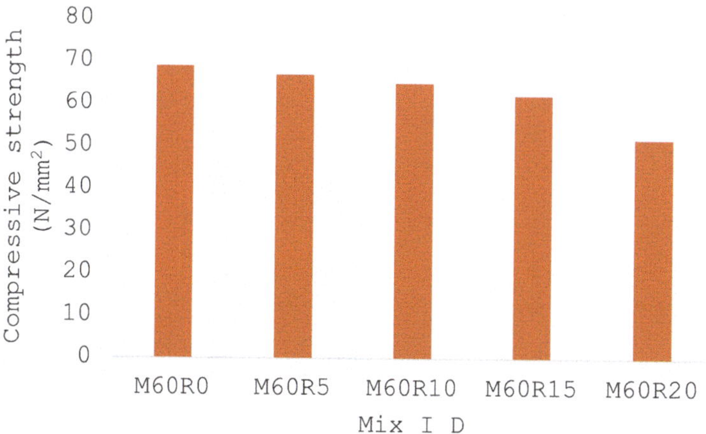

Fig. 2.5 Compressive strength of rubcrete

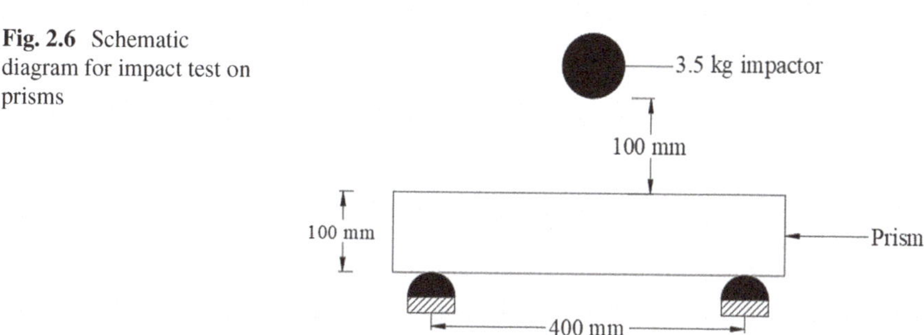

Fig. 2.6 Schematic diagram for impact test on prisms

cost ratio. This ratio balances the impact energy absorption capacity of the concrete with the cost of materials, thereby identifying the most effective composition for real-world structural applications subjected to impact loads. Prisms measuring 100 × 100 × 500 mm, prepared using rubcrete were subjected to a repeated drop-weight impact test. The schematic representation of the test setup is shown in Fig. 2.6. During testing, an effective span of 400 mm was maintained. A drop-weight impactor of 3.5 kg mass was released from a height of 100 mm onto the top surface of the prism specimens. Successive impacts were applied until visible cracks were observed at the bottom face of the specimen. The energy absorption capacity was calculated using Eq. (2.1).

$$U = Nmgh \tag{2.1}$$

where U represents the cumulative energy absorbed in Nm, m is the mass of the impactor in kg, N stands for the number of blows till the appearance of cracks at the bottom of the prism, g is the acceleration due to gravity (9.81 m/s^2), and h signifies the height of fall in m.

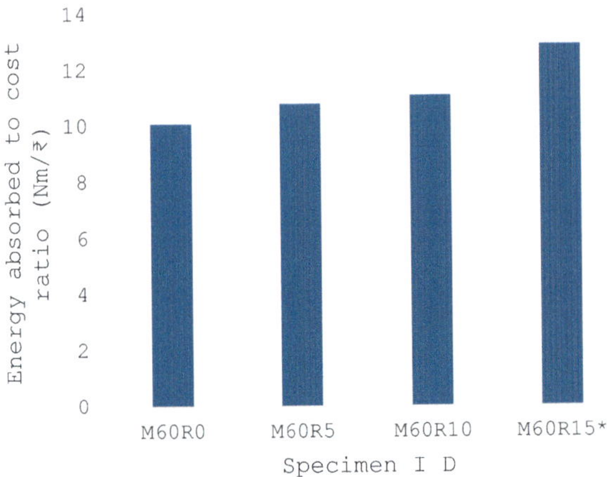

Fig. 2.7 Energy absorbed to cost ratio for rubcrete. (*Optimum rubber content for 60 N/mm^2 concrete = 15% replacement of fine aggregates)

Figure 2.7 presents the details of energy absorbed to cost ratio for rubcrete mixes. The optimum percentage of crumb rubber is 15%. The same percentage of replacement of fine aggregates with crumb rubber was found to be optimum for concretes of strength 20 and 40 N/mm^2.

2.2.6 Fibres

Addition of fibres could compensate the reduction of strength of concrete due to the addition of crumb rubber. They also enhance the impact resistance and energy absorption capacity of concrete. Discrete fibres were introduced into the mix. Two types of fibres were considered: crimped steel fibres (SF) and polypropylene (PP) fibres. These fibres serve distinct roles in controlling cracking and improving toughness. The inclusion of fibres transforms the brittle plain concrete into a quasi-ductile composite, improving its resistance to cracking and dynamic loading. Steel fibres are effective in controlling macrocracks that develop under impact or overload. Their high tensile strength and crimped geometry improve anchorage and load transfer across cracked sections.

Polypropylene fibres, being fine and flexible, are more effective in controlling microcracks. They reduce plastic shrinkage cracking and delay the initiation of cracks under early age and low-level stresses.

Their physical and mechanical properties are summarised in Table 2.6, while Figs. 2.8 and 2.9 show the actual fibre forms used.

Optimum Percentage of Fibres The percentage of steel fibres was varied as 0.25%, 0.5%, 0.75%, and 1% by volume of concrete, whereas percentage of polypropylene fibres was 0.1%, 0.2%, and 0.3%. Determination of optimum proportions of steel fibres in steel fibre-reinforced concrete and polypropylene fibres in polypropylene fibre-reinforced concrete was determined by following the same procedure as in

Table 2.6 Properties of fibres

Property	Steel fibres	Polypropylene fibres
Type	Round crimped	Monofilament
Length	30 mm	12 mm
Aspect ratio	60	–
Diameter	–	0.02 mm
Tensile strength	1100 N/mm^2	551 N/mm^2

Fig. 2.8 Crimped steel fibres

Fig. 2.9 Polypropylene fibres

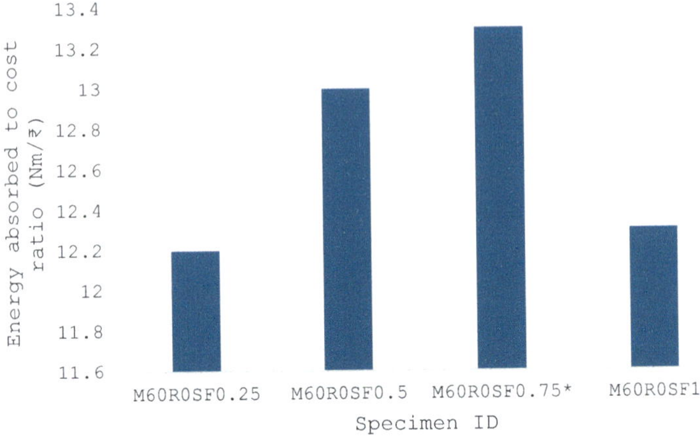

Fig. 2.10 Energy absorbed to cost ratio for 60 N/mm² steel fibre-reinforced concrete. (*Optimum steel fibre content for 60 N/mm² steel fibre-reinforced concrete = 0.75% by volume)

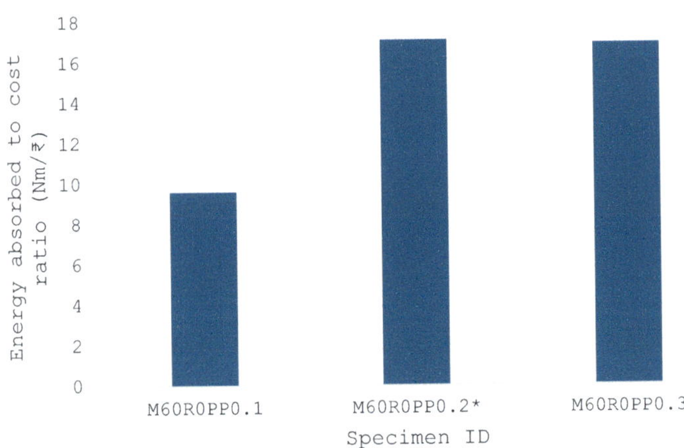

Fig. 2.11 Energy absorbed to cost ratio for 60 N/mm² polypropylene fibre-reinforced concrete. (*Optimum polypropylene fibre content for 60 N/mm² polypropylene fibre-reinforced concrete = 0.2% by volume)

rubcrete. Figures 2.10 and 2.11 contain the results of the energy absorbed to cost ratio of steel fibre-reinforced concrete and polypropylene fibre-reinforced concrete.

Mix Ratio of Different Types of Concrete Tables 2.7, 2.8, 2.9, and 2.10 summarise the details of mix variants and mix proportions of all mixes.

Table 2.7 Details of mix variants

Mix ID	Rubber (%)	Steel fibres (%)	Polypropylene fibres (%)
MGR0	0	0	0
MGR5	5	0	0
MGR10	10	0	0
MGR15	15	0	0
MGR20	20	0	0
MGR0SF0.25	0	0.25	0
MGR0SF0.5	0	0.5	0
MGR0SF0.75	0	0.75	0
MGR0SF1	0	1	0
MGR15SF0.75	15	0.75	0
MGR0PP0.1	0	0	0.1
MGR0PP0.2	0	0	0.2
MGR0PP0.3	0	0	0.3
MGR15PP0.2	15	0	0.2

In Tables 2.7, 2.8, 2.9, and 2.10, MG is the compressive strength of the concrete mix. G can take values of 20, 40, and 60 for concrete with compressive strength 20 N/mm^2, 40 N/mm^2, and 60 N/mm^2 respectively. In RX, X is the percentage of replacement of fine aggregates with crumb rubber. In SFY, Y is the percentage volume of steel fibres in the total volume of concrete. In PPZ, Z is the percentage volume of polypropylene fibres in the total volume of concrete

Table 2.8 Mix proportion for concrete of strength 20 N/mm^2

Mix ID	Cem	FA	CA	Wat	Rub	SF	PP	Slump (mm)
M20R0	1.000	1.808	3.589	0.481	0.000	0.000	0.000	130
M20R5	1.000	1.758	3.589	0.481	0.022	0.000	0.000	125
M20R10	1.000	1.664	3.589	0.481	0.044	0.000	0.000	120
M20R15	1.000	1.572	3.589	0.481	0.067	0.000	0.000	115
M20R20	1.000	1.481	3.589	0.481	0.089	0.000	0.000	90
M20R0SF0.25	1.000	1.808	3.591	0.479	0.000	0.056	0.000	115
M20R0SF0.5	1.000	1.810	3.592	0.480	0.000	0.109	0.000	105
M20R0SF0.75	1.000	1.810	3.594	0.482	0.000	0.165	0.000	90
M20R0SF1	1.000	1.809	3.593	0.480	0.000	0.222	0.000	70
M20R15SF0.75	1.000	1.574	3.594	0.482	0.067	0.165	0.000	90
M20R0PP0.1	1.000	1.806	3.586	0.481	0.000	0.000	0.003	110
M20R0PP0.2	1.000	1.811	3.593	0.479	0.000	0.000	0.006	85
M20R0PP0.3	1.000	1.808	3.588	0.479	0.000	0.000	0.008	65
M20R15PP0.2	1.000	1.574	3.593	0.479	0.067	0.000	0.006	80

In Tables 2.8, 2.9, and 2.10, *Cem*- cement; *FA*- fine aggregates; *CA*- coarse aggregates; *Wat*- water; *SP*- superplasticiser; *SP*- superplasticiser; *SF*- steel fibres; *PP*- polypropylene fibres; *HR*- high-range water reducer; *VM*- viscosity modifying agent

Table 2.9 Mix proportion for concrete of strength 40 N/mm^2

Mix ID	Cem	FA	CA	Wat	SP	Rub	SF	PP	Slump (mm)
M40R0	1.000	1.857	2.221	0.400	0.007	0.000	0.000	0.000	117
M40R5	1.000	1.806	2.221	0.400	0.007	0.022	0.000	0.000	114
M40R10	1.000	1.710	2.221	0.400	0.007	0.047	0.000	0.000	110
M40R15	1.000	1.616	2.221	0.400	0.007	0.069	0.000	0.000	105
M40R20	1.000	1.520	2.221	0.400	0.007	0.092	0.000	0.000	80
M40R0SF0.25	1.000	1.857	2.221	0.400	0.007	0.000	0.045	0.000	104
M40R0SF0.5	1.000	1.857	2.220	0.401	0.007	0.000	0.087	0.000	95
M40R0SF0.75	1.000	1.856	2.220	0.400	0.007	0.000	0.133	0.000	79
M40R0SF1	1.000	1.856	2.221	0.401	0.007	0.000	0.178	0.000	55
M40R15SF0.75	1.000	1.613	2.220	0.400	0.007	0.070	0.133	0.000	75
M40R0PP0.1	1.000	1.857	2.219	0.400	0.007	0.000	0.000	0.002	96
M40R0PP0.2	1.000	1.859	2.221	0.400	0.007	0.000	0.000	0.004	73
M40R0PP0.3	1.000	1.857	2.219	0.400	0.007	0.000	0.000	0.007	52
M40R15PP0.2	1.000	1.615	2.221	0.400	0.007	0.069	0.000	0.004	68

Table 2.10 Mix proportion for concrete of strength 60 N/mm^2

Mix ID	Cem	Met	FA	CA	Wat	HR	VM	Rub	SF	PF	Slump (mm)
M60R0	1.000	0.093	1.519	2.701	0.367	0.009	0.005	0.000	0.000	0.000	105
M60R5	1.000	0.093	1.442	2.701	0.367	0.009	0.005	0.018	0.000	0.000	100
M60R10	1.000	0.093	1.367	2.701	0.367	0.009	0.005	0.036	0.000	0.000	93
M60R15	1.000	0.093	1.290	2.701	0.367	0.009	0.005	0.054	0.000	0.000	86
M60R20	1.000	0.093	1.215	2.701	0.367	0.009	0.005	0.075	0.000	0.000	78
M60R0SF0.25	1.000	0.093	1.518	2.702	0.366	0.009	0.005	0.000	0.045	0.000	92
M60R0SF0.5	1.000	0.091	1.522	2.706	0.367	0.009	0.005	0.000	0.089	0.000	85
M60R0SF0.75	1.000	0.091	1.523	2.710	0.368	0.009	0.005	0.000	0.135	0.000	78
M60R0SF1	1.000	0.092	1.524	2.712	0.366	0.009	0.005	0.000	0.181	0.000	52
M60R15SF0.75	1.000	0.091	1.295	2.710	0.368	0.009	0.005	0.055	0.135	0.000	69
M60R0PP0.1	1.000	0.091	1.519	2.701	0.367	0.009	0.005	0.000	0.000	0.002	80
M60R0PP0.2	1.000	0.091	1.520	2.705	0.368	0.009	0.005	0.000	0.000	0.005	70
M60R0PP0.3	1.000	0.091	1.520	2.702	0.366	0.009	0.005	0.000	0.000	0.007	49
M60R15PP0.2	1.000	0.091	1.293	2.705	0.368	0.009	0.005	0.055	0.000	0.005	65

Optimum Mix Proportions The optimum proportions of crumb rubber, steel fibres, and polypropylene fibres were found to be the same for all strengths of concrete (20, 40, and 60 N/mm^2). Among the rubcrete mixes, the highest energy absorption-to-cost ratio was observed in the mix where 15% of fine aggregates were replaced by an equivalent volume of crumb rubber. Optimal fibre dosages were identified as 0.75% for steel fibres in steel fibre-reinforced concrete and 0.2% for polypropylene fibres in polypropylene fibre-reinforced concrete. Consequently, subsequent studies were conducted using these ideal proportions of crumb rubber, steel fibres, and polypropylene fibres.

2.3 Fresh Properties

The fresh properties of concrete reflect its ease of placement, compaction, and finishing. Slump tests were conducted to assess the workability of the various mixes (Tables 2.8, 2.9, and 2.10). The inclusion of crumb rubber was observed to reduce slump values, primarily due to a reduction in internal cohesion within the concrete matrix. The angular and elastic nature of rubber particles tends to disrupt the homogeneity of the fresh mix. Similarly, the incorporation of steel fibres significantly increased the stiffness of the mix, further reducing workability. In particular, when the steel fibre content reached 1%, a noticeable balling effect occurred, hindering uniform distribution. Polypropylene fibres also contributed to a decline in workability, as their high surface area and fibrous nature tend to absorb more mixing water and increase internal friction.

2.4 Engineering Properties

This section presents the mechanical behaviour of rubcrete and its fibre-reinforced variants. The focus is on compressive strength, stress–strain relationship, flexural strength, split tensile strength, and fracture energy. These properties were evaluated for the optimised mixes identified in the previous section.

The following concrete types were tested across minimum strength of 20, 40, and 60 N/mm^2. Ordinary concrete, rubcrete with 15% crumb rubber, Steel fibre-reinforced concrete (SFRC) with 0.75% crimped steel fibres, Steel fibre-reinforced rubcrete (SFRR) with 15% crumb rubber and 0.75% steel fibres, polypropylene fibre-reinforced concrete (PFRC) with 0.2% polypropylene fibres, and polypropylene fibre-reinforced rubcrete (PFRR) with 15% crumb rubber, and 0.2% polypropylene fibres.

2.4.1 Compressive Strength

Compressive strength is one of the most fundamental mechanical properties of concrete. It reflects the material's ability to resist axial loads and provides a basic measure of overall performance and quality. The 28-day compressive strength of rubcrete and its fibre-reinforced variants was evaluated to understand the influence of crumb rubber, steel fibres, and polypropylene fibres on strength development.

The compressive strength of the various mixes was evaluated in accordance with IS 516: 1959 (reaffirmed in 2004) [6] using 150 mm cube specimens. The variation in compressive strength across the fibre-reinforced rubcrete mixes is illustrated in Fig. 2.12.

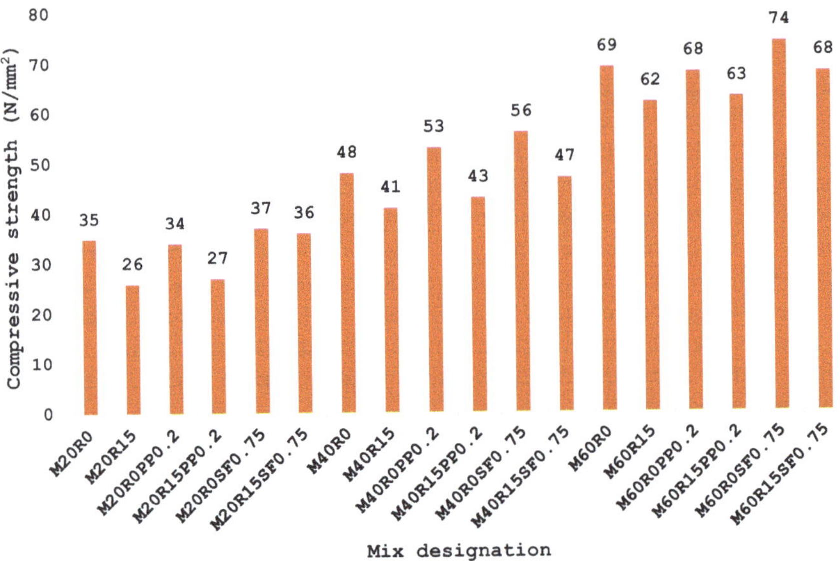

Fig. 2.12 Compressive strength of mixes

Across all grades of concrete, the inclusion of 15% crumb rubber led to a noticeable reduction in compressive strength, primarily due to the lower stiffness and poor bonding characteristics of rubber particles. However, this strength loss was partially compensated by the addition of fibres. Steel fibres had a more pronounced effect in improving compressive strength than polypropylene fibres. In 20 and 40 N/mm^2 concrete, steel fibre-reinforced rubcrete (SFRR) restored the strength nearly to the strength of ordinary concrete. In 60 N/mm^2, where high-strength matrix and denser microstructure dominate, both fibre types improved performance, with SFRR reaching over 97% of the strength of ordinary concrete. The results demonstrate that fibres mitigate the strength reduction associated with rubber, making fibre-reinforced rubcrete a viable material for structural applications that demand moderate to high compressive strength.

2.4.2 *Stress–Strain Relationship*

The stress–strain relationship provides vital insights into the deformational behaviour and ductility of concrete. While compressive strength indicates the load-bearing capacity, the stress–strain curve reveals how concrete responds under progressive loading, including its elastic limit, strain capacity, and energy absorption before failure. Cylindrical specimens of size 150 mm diameter and 300 mm height were tested using a displacement-controlled universal testing machine, equipped with LVDTs to measure axial deformation. The test was conducted at a loading rate of

0.2 mm/min. Stress–strain curves were plotted for each mix to evaluate peak strain, post-peak behaviour, and strain energy density (area under the curve). Figures 2.13, 2.14, and 2.15 present the stress–strain curves of the mixes [5].

Ordinary concrete showed a typical brittle response with a well-defined peak followed by a sharp drop in stress. Rubcrete (RC) exhibited a gentler post-peak

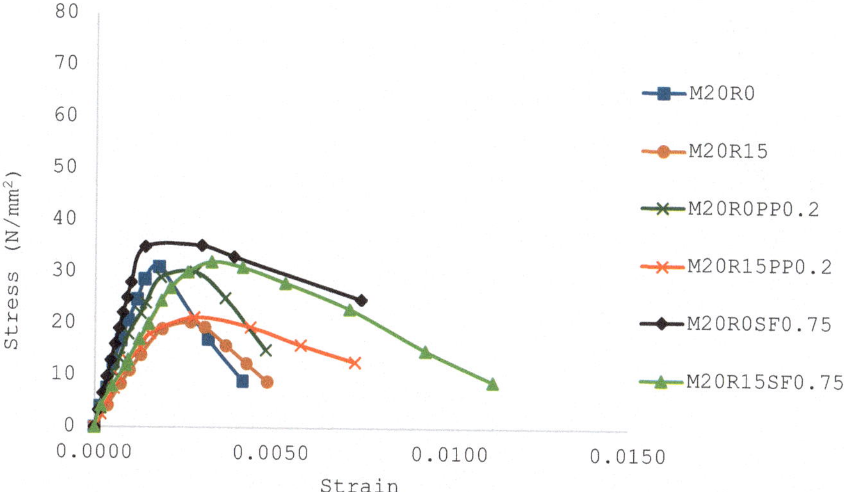

Fig. 2.13 Stress–Strain curves for concrete of strength 20 N/mm²

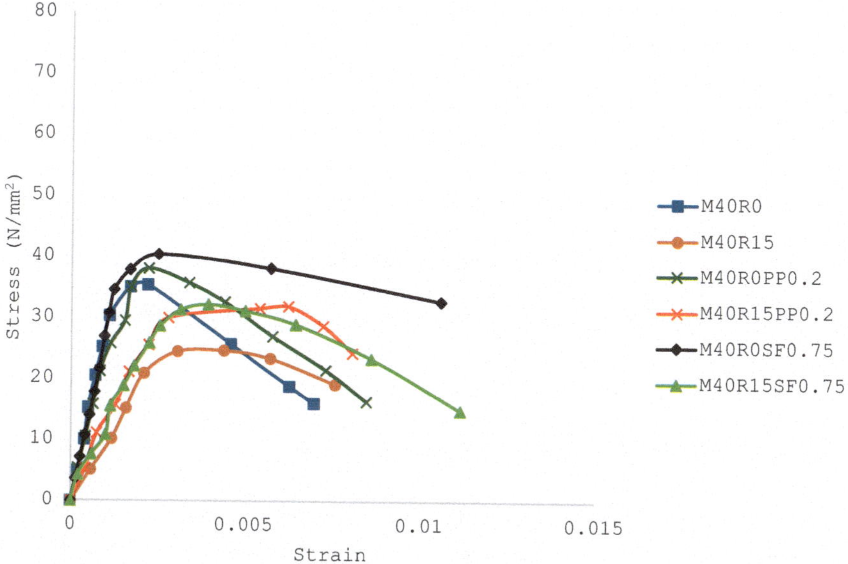

Fig. 2.14 Stress–Strain curves for concrete of strength 40 N/mm²

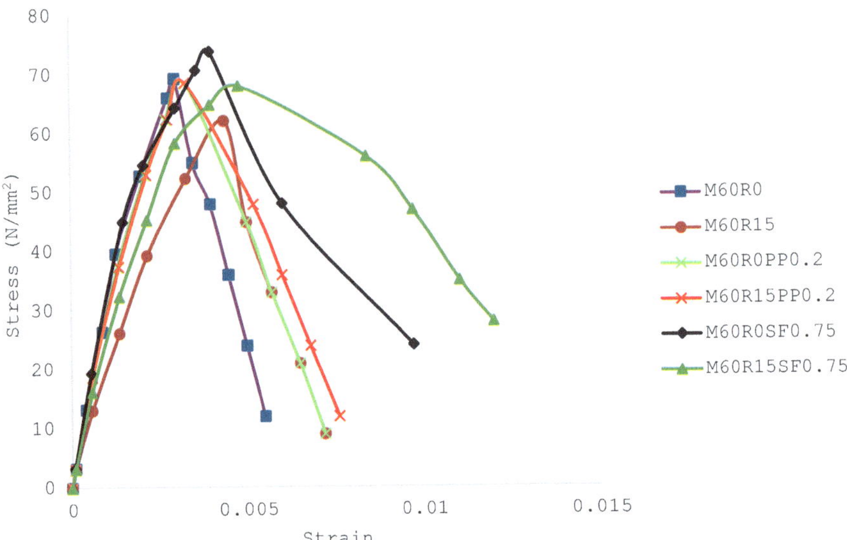

Fig. 2.15 Stress–Strain curves for concrete of strength 60 N/mm^2

slope due to the elastic deformation of rubber particles, indicating better energy dissipation but lower peak strength. SFRC and SFRR demonstrated higher peak strains and a gradual descending branch, signifying improved ductility and toughness. Polypropylene fibres in PFRC and PFRR contributed to strain redistribution and crack-bridging, resulting in a modestly enhanced post-peak behaviour, although less pronounced than steel fibres.

Inclusion of rubber improves the strain energy density and post-peak deformation, while fibres enhance both peak strain and energy absorption. The combination of crumb rubber and steel fibres in SFRR provided the most balanced behaviour, achieving both strength and ductility. The area under the stress–strain curve representing strain energy density was significantly higher for SFRC and SFRR mixes, underlining their superior toughness. Polypropylene fibres also contributed to crack control and ductility but to a lesser extent. These findings reinforce the potential of fibre-reinforced rubcrete in impact-resistant and energy-dissipating applications, particularly in structures subjected to dynamic and accidental loads.

2.4.3 *Flexural Strength*

Flexural strength represents the ability of a concrete element to resist bending or tensile stresses across its depth. It is particularly relevant for structures such as beams, pavements, kerbs, fencing posts, and sleepers, where flexural tension governs failure. In fibre-reinforced systems, flexural behaviour is also

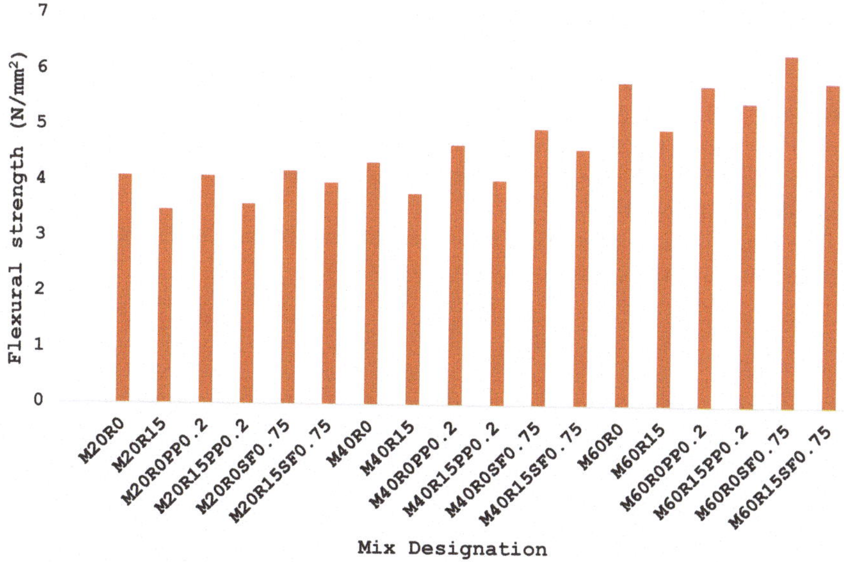

Fig. 2.16 Flexural strength of mixes

indicative of post-cracking resistance and ductility. To determine flexural strength, 100 mm × 100 mm × 500 mm prism specimens were cast and tested under two-point loading [6]. All specimens were tested at 28 days of curing. Figure 2.16 shows the flexural strength of the mixes.

The incorporation of crumb rubber in concrete led to a reduction in flexural strength across all grades of concrete due to the lower stiffness and weak interfacial bonding of rubber particles. However, the addition of fibres effectively mitigated this reduction. Steel fibres were more effective in enhancing flexural strength than polypropylene fibres due to their higher stiffness and tensile strength, as well as their ability to bridge macrocracks. Polypropylene fibres, though less stiff, contributed to improved pre-cracking behaviour and offered marginal post-crack resistance. In rubcrete mixes, both fibre types improved strength, but steel fibres consistently outperformed polypropylene fibres in restoring the flexural strength of control mixes. These findings suggest that fibre reinforcement can effectively restore and enhance flexural performance, making rubcrete suitable for structural elements subjected to flexural stresses and impact forces.

2.4.4 Fracture Energy

Fracture energy represents the energy required to propagate a crack through a unit area of concrete. It is a critical parameter for materials subjected to dynamic loads and impact, as it reflects the concrete's resistance to cracking and brittleness. In fibre-reinforced systems, it serves as a reliable measure of toughness and post-crack ductility.

To evaluate fracture energy, notched beam specimens of size 100 mm × 100 mm × 500 mm were prepared with a central notch of 5 mm width and 30 mm depth. Testing was performed under three-point bending, in accordance with RILEM TC 50-FMC [7] recommendations, using a closed-loop servo-controlled testing machine.

The load–displacement response was recorded, and the area under the curve was used to compute the fracture energy using Eq. (2.2):

$$G_f = \frac{W_0 + mg\delta_0}{A_{lig}} \tag{2.2}$$

where G_f is the fracture energy in Nm/m^2, W_0 is the area under the load-deflection graph in Nm, mg is the self-weight of the beam between the supports in N, δ_0 is the deformation of the beam at failure in m, A_{lig} is the projection of the fracture zone on a plane perpendicular to the beam axis in m^2.

The fracture energy of prisms is presented in Fig. 2.17. The inclusion of crumb rubber led to a noticeable increase in fracture energy compared to control mixes, despite a reduction in compressive strength. This is attributed to the elastic deformation of rubber particles, which absorb and redistribute energy during crack propagation.

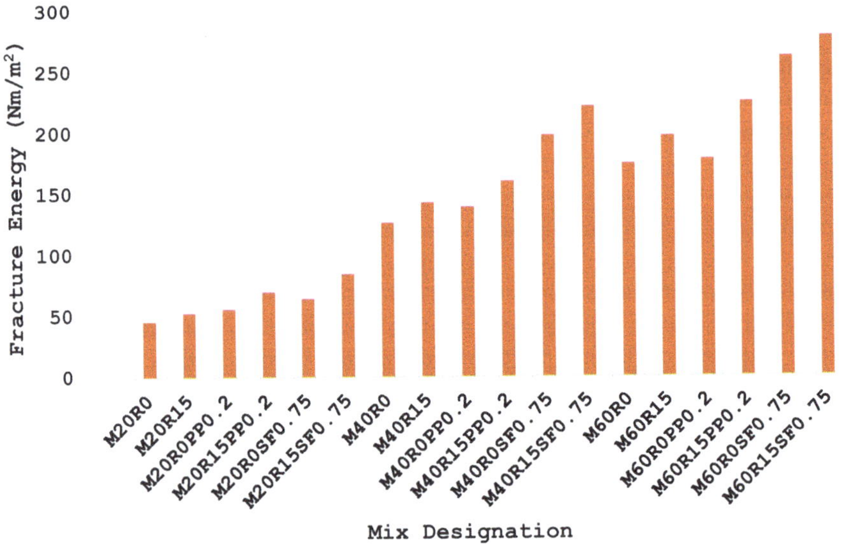

Fig. 2.17 Fracture energy of prisms

- Steel fibre-reinforced mixes (SFRC and SFRR) exhibited the highest fracture energy, owing to the crack-bridging action and high tensile strength of steel fibres. The combined presence of rubber and steel fibres in SFRR yielded a significant improvement in energy dissipation.
- Polypropylene fibres also enhanced fracture energy, though to a lesser extent. Their ability to control microcracks and distribute strain improved the ductile response, especially in PFRR mixes.

More details on the fracture parameter studies of rubcrete can be found in the following references [8–10]:

Summary of Mechanical Properties
This section evaluated the mechanical performance of rubcrete and its fibre-reinforced variants for 20, 40, and 60 N/mm^2 concrete. Key parameters considered were compressive strength, stress–strain behaviour, flexural strength, and fracture energy.

- The incorporation of crumb rubber alone led to a moderate reduction in compressive and flexural strength, attributed to the softer and less stiff nature of rubber particles. However, it significantly enhanced strain capacity and fracture energy, indicating improved energy absorption and post-crack behaviour.
- The inclusion of steel fibres substantially improved both strength and ductility. Concrete with 0.75% crimped steel fibres (SFRC) showed enhanced compressive and flexural strength, a more ductile stress–strain response, and the highest fracture energy among all mixes. Steel fibre-reinforced rubcrete (SFRR) combined the benefits of both rubber and fibres, offering the best energy absorption performance.
- Polypropylene fibres, though not as strong as steel fibres, provided notable improvements in fracture energy and post-peak behaviour. Polypropylene fibre-reinforced rubcrete (PFRR) demonstrated improved toughness and resistance to crack propagation, especially in the early stages of loading.

Overall, the mechanical test results confirm that rubcrete, when optimally reinforced with fibres, can offer a balanced performance, combining sustainability, toughness, and acceptable strength for use in structures subjected to impact and dynamic loading. The superior performance of SFRR makes it a promising candidate for applications like sleepers, crash barriers, and fencing posts.

2.5 Durability

Durability is a critical performance parameter for concrete structures, especially those exposed to aggressive environments over long service lives. For structural elements such as fencing posts, railway sleepers, and crash barriers often subjected to cycles of wetting and drying, chemical exposure, and temperature fluctuations, the ability of concrete to resist degradation is vital.

In this section, the durability performance of rubcrete and its fibre-reinforced variants is evaluated through the following tests:

- Water absorption.
- Sorptivity.
- Resistance to sulphate attack
- Resistance to acid attack
- Resistance to marine water exposure

These tests were carried out on 40 and 60 N/mm^2 rubcrete specimens. The results offer insights into how the inclusion of crumb rubber, steel fibres, and polypropylene fibres affects the permeability and chemical resistance of concrete mixes. All tests were performed on 28-day cured specimens, following relevant testing standards.

2.5.1 Water Absorption

Water absorption is a key indicator of the permeability and pore structure of concrete. High water absorption suggests a more porous microstructure, which can compromise the concrete's resistance to weathering, chemical ingress, and freeze–thaw damage [11]. Water absorption was determined based on IS 1124: 1974 (Reaffirmed 2003) [11, 12].

The 100 mm cubes were first cured, then surface-dried and placed in an oven for 24 h to ensure complete drying. After oven drying, their weights were recorded as W_1. Subsequently, the specimens were immersed in water for 24 h, after which the saturated weights (W_2) were measured. Water absorption was calculated using the following Eq. (2.3):

$$\text{Percentage of water absorbed} = \frac{(W_2 - W_1) \times 100}{W_1} \quad (2.3)$$

Figure 2.18 illustrates the percentage of water absorbed by 100 mm concrete cubes. According to the International Federation for Structural Concrete [13], concrete exhibiting early water absorption below 3% is considered to possess good durability. As shown in the figure, all tested specimens recorded water absorption values below this threshold, indicating satisfactory performance in terms of permeability.

2.5.2 Sorptivity

Sorptivity is a measure of the capillary suction of water into unsaturated concrete. It reflects the rate at which water can be drawn into the pore system and is a critical indicator of concrete's resistance to moisture ingress, which can lead to

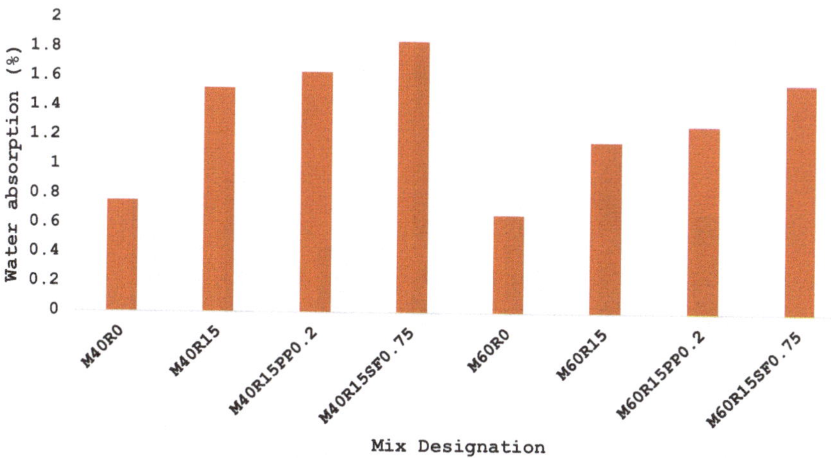

Fig. 2.18 Results of water absorption test

deterioration through freeze–thaw cycles, corrosion of reinforcement, or chemical attack. The sorptivity was measured according to the method outlined in ASTM C1585–2013 [14]. Cylindrical specimens of 100 mm diameter × 50 mm height were dried in an oven at 50 °C for 72 h and cooled to room temperature. The lateral surface of the specimen was sealed with epoxy resin, leaving only one face exposed. The specimen was placed in a shallow water tray with 3–5 mm depth of water in contact with the exposed face. The change in weight due to water absorption was measured at regular time intervals up to 6 h. Sorptivity was calculated using Eqs. (2.4) and (2.5). The results of sorptivity test are shown in Fig. 2.19.

$$S = \frac{I}{t^{\frac{1}{2}}} \tag{2.4}$$

$$I = \frac{\Delta W}{Ad} \tag{2.5}$$

where S is the sorptivity in mm and t stands for the elapsed time in minutes. ΔW is the change in weight of cylinders in g. A is the surface area of the specimen through which water penetrates in mm^2, and d is the density of water in g/mm^3.

The sorptivity values of all tested specimens were found to be below the maximum permissible limit of 9×10^{-2} mm/min$^{0.5}$. While the incorporation of rubber and fibres led to a nominal increase in moisture ingress, all mix variants, namely ordinary concrete, rubcrete, steel fibre-reinforced rubcrete, and polypropylene fibre-reinforced rubcrete remained within the acceptable range, demonstrating satisfactory resistance to capillary absorption.

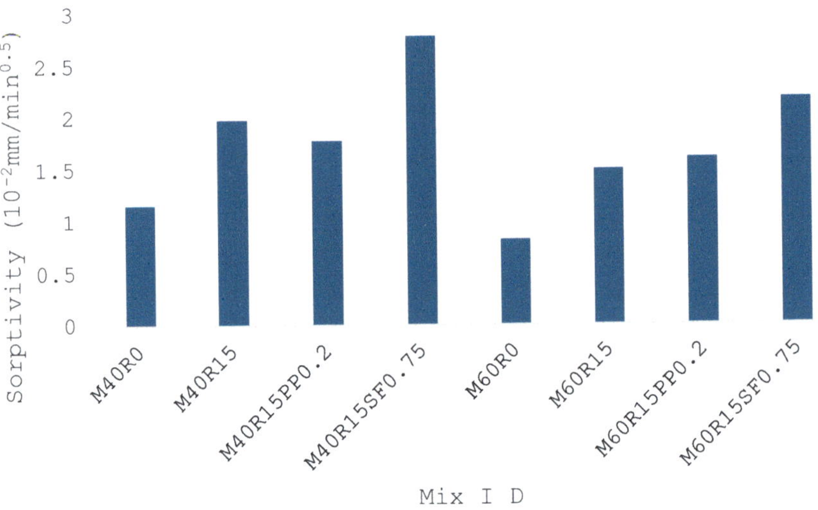

Fig. 2.19 Sorptivity of specimen

2.5.3 *Resistance to Sulphate Attack*

Concrete structures exposed to sulphate-rich environments such as soils, groundwater, and industrial effluents, are at risk of deterioration due to sulphate attack. This chemical reaction primarily affects the hydrated cement paste, leading to expansion, cracking, and loss of strength. Evaluating the resistance of rubcrete and its fibre-reinforced variants to sulphate exposure is essential for understanding their long-term durability in aggressive environments. The testing is done by immersing 100 mm × 100 mm × 100 mm cube specimens in a 30 g/L sodium sulphate (Na_2SO_4) solution for 90 days. Specimens were initially water-cured for 28 days prior to immersion. Visual inspection was performed to assess surface deterioration, cracking, or expansion. Residual compressive strength was determined after sulphate exposure and compared to control specimens stored in water. The results are presented in Table 2.11.

2.5.4 *Resistance to Acid Attack*

Concrete is particularly vulnerable to acidic environments, where hydrogen ions aggressively react with the alkaline components of cement paste. This results in leaching, matrix disintegration, and surface erosion, compromising both strength and durability. This section evaluates the acid resistance of rubcrete and its fibre-reinforced variants, simulating exposure conditions commonly encountered in industrial or sewage-related environments.

Table 2.11 Results of sulphate attack tests

Specimen ID	Weight of cube (kg)	Weight of cube after 90 days (kg)	Percent change in weight	Compressive strength (N/ mm^2)	Compressive strength after 90 days (N/ mm^2)	Percentage reduction in strength
M40R0	2.454	2.449	0.2	50	40	20
M40R15	2.377	2.372	0.21	41	34	17.07
M40R15PP0.2	2.375	2.37	0.21	43	36	16.28
M40R15SF0.75	2.393	2.387	0.25	47	39	17.02
M60R0	2.51	2.505	0.2	61	50	18.03
M60R15	2.43	2.426	0.16	52	44	15.38
M60R15PP0.2	2.432	2.428	0.16	54	45	16.67
M60R15SF0.75	2.482	2.477	0.2	64	53	17.19

Table 2.12 Results of acid attack tests

Specimen ID	Weight of cube (kg)	Weight of cube after 90 days (kg)	Percentage change in weight	Compressive strength (N/mm^2)	Compressive strength after 90 days (N/mm^2)	Percentage reduction in strength
M40R0	2.454	2.171	11.53	50	36	28.00
M40R15	2.377	2.214	6.86	41	32	21.95
M40R15PP0.2	2.375	2.194	7.62	43	33	23.26
M40R15SF0.75	2.393	2.141	10.53	47	36	23.40
M60R0	2.510	2.272	9.48	61	47	22.95
M60R15	2.430	2.296	5.51	52	43	17.31
M60R15PP0.2	2.432	2.266	6.83	54	44	18.52
M60R15SF0.75	2.482	2.260	8.94	64	52	18.75

100 mm cube specimens were cured in water for 28 days. The specimens were then immersed in a 3% H_2SO_4 solution for 90 days. Visual examination was conducted periodically for signs of deterioration. After 90 days, specimens were tested for residual compressive strength, and results were compared to identical specimens stored in water. The results are given in Table 2.12.

The acid exposure led to a noticeable reduction in compressive strength for all mixes.

Resistance to Marine Water Exposure

Concrete elements exposed to marine environments must withstand a combination of chloride ingress, sulphate presence, and moisture cycles, all of which can significantly deteriorate the concrete matrix and corrode embedded reinforcement. Evaluating rubcrete and its fibre-reinforced variants under marine water conditions is vital for assessing their suitability in coastal and offshore infrastructure. 100 mm cube specimens were cured in water for 28 days. They were then immersed in natural marine water produced as per as per ASTM D 1141 [15] for 90 days. At the end of exposure, specimens were tested for residual compressive strength. Visual observations were also recorded to assess surface deterioration. Figure 2.20 and Table 2.13 show the composition of marine water and results of marine water exposure test.

Summary of Durability Tests

The durability performance of rubcrete and its fibre-reinforced variants was systematically evaluated through water absorption, sorptivity, and resistance to sulphate, acid, and marine water exposure. While the inclusion of crumb rubber increased the permeability of the concrete matrix evident from elevated water absorption and sorptivity values it did not lead to catastrophic deterioration. The impact of rubber-induced porosity was mitigated by the incorporation of steel and polypropylene fibres, both of which contributed to matrix densification and crack control.

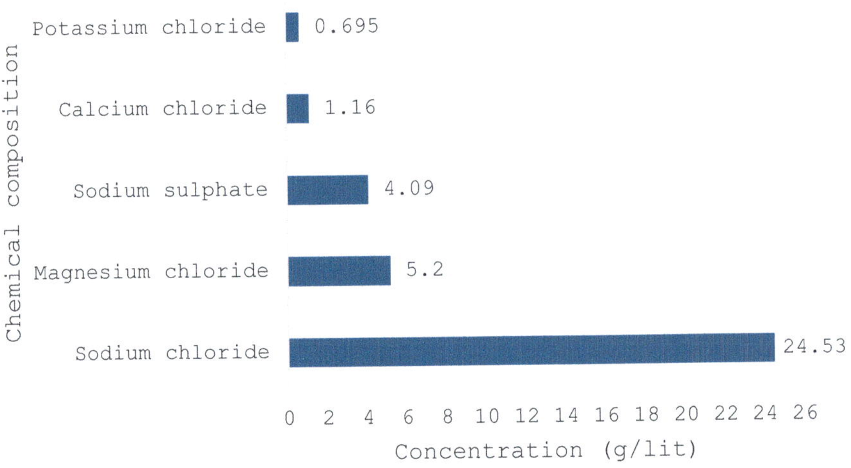

Fig. 2.20 Composition of marine water

Table 2.13 Results of marine water exposure tests

Specimen ID	Weight of cube (kg)	Weight of cube after 90 days (kg)	Percentage change in weight	Compressive strength (N/mm^2)	Compressive strength after 90 days (N/mm^2)	Percentage reduction in strength
M40R0	2.454	2.415	1.59	50	40	20.00
M40R15	2.377	2.365	0.50	41	37	9.76
M40R15PP0.2	2.375	2.360	0.63	43	38	11.63
M40R15SF0.75	2.393	2.372	0.88	47	40	14.89
M60R0	2.510	2.473	1.47	61	52	14.75
M60R15	2.430	2.421	0.37	52	48	7.69
M60R15PP0.2	2.432	2.422	0.41	54	49	9.26
M60R15SF0.75	2.482	2.468	0.56	64	58	9.38

Across all exposure conditions, steel fibre-reinforced rubcrete (SFRR) exhibited superior resistance, maintaining higher residual strengths and showing no signs of surface degradation. Polypropylene fibre-reinforced rubcrete (PFRR) also performed well, particularly in reducing microcracking and retaining strength in acid and marine environments.

These results affirm that, although rubberised concrete poses inherent durability challenges, the combined use of discontinuous fibres and surface-modified crumb rubber can yield a durable and sustainable material suitable for structures exposed to aggressive conditions.

2.6 Summary

This chapter presented the development and characterisation of fibre-reinforced rubcrete, focusing on its material selection, mix proportioning, and evaluation of fresh, mechanical, and durability properties for three strength grades: 20, 40, and 60 N/mm^2. The chapter began by discussing the functional roles of each constituent material: cement, metakaolin, fine and coarse aggregates, crumb rubber, fibres, and admixtures with emphasis on their influence on performance.

Optimum rubcrete mixes were developed by replacing 15% of the volume of fine aggregates with pretreated crumb rubber and reinforcing them with different percentages of steel fibres and polypropylene fibres. A detailed mixing methodology was employed to ensure homogeneity and uniform fibre dispersion. The fresh properties of the mixes were assessed, followed by a cost-to-energy absorption analysis that guided the selection of optimum fibre and crumb rubber content (0.75% for steel fibres and 0.2% for polypropylene fibre).

The mechanical performance of the optimised mixes was examined through compressive strength, flexural strength, stress–strain response, and fracture energy. It was observed that rubcrete showed a reduction in strength properties compared to ordinary concrete but demonstrated significantly enhanced strength and energy absorption capacity when fibres were incorporated. Steel fibre-reinforced rubcrete exhibited the best performance in terms of strength retention and toughness.

Durability was assessed through water absorption, sorptivity, and exposure to sulphate, acid, and marine environments. While rubber inclusion increased permeability, the addition of fibres helped mitigate adverse effects, particularly in harsh conditions.

In conclusion, this chapter established the feasibility of developing structurally sound, impact-resistant, and moderately durable concrete using crumb rubber and fibres. These mixes will now be used in structural elements such as prestressed railway sleepers, fencing posts, crash barriers, and kerbs in the chapters that follow.

For further details on the development of fibre-reinforced rubcrete and its properties, the readers could also refer the journals published by the authors [8, 11, 16–19].

References

1. Bureau of Indian Standards, Portland Pozzolana Cement-specification. IS 1489 (Part 1): 1991, New Delhi, India (1991)
2. Bureau of Indian Standards, Indian standard ORDINARY PORTLAND CEMENT, 53 GRADE—SPECIFICATION (first revision) IS. IS 12269: 2013, New Delhi, India (2013)
3. Bureau of Indian Standards, Coarse and fine aggregate for concrete—specification (third revision). IS 383: 2016 third edit, New Delhi, India (2016)
4. Bureau of Indian Standards, Concrete, Plain and Reinforced. IS. 456: 2000 (Reaffirmed 2005), New Delhi, India (2000)
5. A. Raj, P. Nagarajan, A.P. Shashikala, *Application of Fibre Reinforced Rubcrete in Structures Subjected to Impact Load* (National Institute of Technology Calicut, 2021)
6. Bureau of Indian Standards, Method of tests for strength of concrete. IS: 516-1959 (reaffirmed 2004), New Delhi, India (2004)
7. RILEM TC-50 FMC, Determination of the fracture energy of mortar and concrete by means of three-point bend tests on notched beams. Mater. Struct./Mater. Constr. **8**, 285–296 (1985). https://doi.org/10.1007/BF00962380
8. A. Raj, P.J. Usman Arshad, P. Nagarajan, A.P. Shashikala, Fracture behaviour of fibre reinforced rubcrete. Mater. Sci. Forum **969**, 80–85 (2019). https://doi.org/10.4028/www.scientific.net/MSF.969.80
9. A. Raj, P.J. Usman Arshad, P. Nagarajan, A.P. Shashikala, Fracture behaviour of steel fibre reinforced rubcrete, in *Lecture Notes in Civil Engineering*, (Springer, Singapore, 2021), pp. 619–628
10. A. Raj, P.J. Usman Arshad, P. Nagarajan, A.P. Shashikala, Experimental investigation on the fracture behaviour of polypropylene fibre-reinforced rubcrete, in *Lecture Notes in Mechanical Engineering*, (Springer, Singapore, 2020), pp. 335–345
11. A. Raj, P. Nagarajan, A.P. Shashikala, Behaviour of fibre-reinforced rubcrete beams subjected to impact loading. J. Inst. Eng. (India) A (2020). https://doi.org/10.1007/s40030-020-00470-4
12. Bureau of Indian Standards, Method of test for determination of water absorption, apparent specific gravity and porosity of natural building stones. IS: 1124-1974 (Reaffirmed 2003), New Delhi, India (1974)
13. J.C. Walraven, *Model Code 2010-Final draft. fib Fédération internationale du béton* (Lausanne, 2012)
14. ASTM International, Standard test method for measurement of rate of absorption of water by hydraulic-cement concretes. West Conshohocken, USA (2013)
15. American Society for Testing and Materials, Standard Practice for the Preparation of Substitute Ocean Water. West Conshohocken, USA (2013)
16. A. Raj, P. Nagarajan, S. Aikot Pallikkara, Application of Fiber-reinforced rubcrete for crash barriers. J. Mater. Civ. Eng. **32**, 04020358 (2020). https://doi.org/10.1061/(ASCE)MT.1943-5533.0003454
17. A. Raj, P. Nagarajan, S. Aikot Pallikkara, Application of Fiber-reinforced rubcrete in fencing posts. Pract. Period. Struct. Des. Constr. **25**, 04020037 (2020). https://doi.org/10.1061/(asce)sc.1943-5576.0000512
18. A. Raj, P. Nagarajan, A.P. Shashikala, Investigations on fiber-reinforced rubcrete for railway sleepers. ACI Struct. J. **117**, 109–120 (2020). https://doi.org/10.14359/51724679
19. A. Raj, P.J.U. Arshad, P. Nagarajan, A.P. Shashikala, Experimental investigation on the fracture behaviour of polypropylene fibre-reinforced rubcrete, in *Structural Integrity Assessment*, Lecture Notes in Mechanical Engineering, ed. by R. Prakash, K.R. Suresh, A. Nagesha, G. Sasikala, A. Bhaduri, (Springer, Singapore, 2020)

Chapter 3
Railway Sleepers

3.1 Sleepers

Railways have long served as a backbone for the transportation of people and goods across vast distances, fulfilling this role with remarkable efficiency. Among the many components of a railway system, sleepers form a critical element [1, 2]. These structural members are responsible for transferring loads from the train wheels to the ballast layer, which subsequently distributes them to the underlying subgrade [3]. In addition to load transfer, sleepers are essential in preserving both the vertical and horizontal alignment of railway tracks [4]. During derailments, railway sleepers are exposed to high-magnitude impact loads. A representative cross-section of a railway track is shown in Fig. 3.1.

Prestressed concrete has been widely used in modern-day sleepers [5]. The major modes of failure in prestressed concrete sleepers are rail seat abrasion, centre-bound cracking, and high-magnitude impact forces during derailments [6]. Of these, derailments often result in irreparable damage to the sleepers, compromising track integrity and causing substantial delays in railway operations. To minimise the frequency of sleeper replacement and enhance the resilience of rail infrastructure, improving the energy absorption capacity of prestressed concrete sleepers is of paramount importance.

These dynamic demands placed on railway infrastructure, particularly on prestressed concrete sleepers, necessitate innovations in both material composition and structural design. In the search for more resilient and sustainable alternatives, the performance of fibre-reinforced rubcrete and its variants in the context of railway sleeper applications were investigated.

When assessing the performance of the sleeper prototypes, it is essential to understand the minimum performance requirements when the sleepers are subjected to different tests. The minimum ultimate loads for prestressed concrete sleepers as

A. Raj et al., *Rubcrete and Its Applications in Structures Subjected to Impact Loading*, SpringerBriefs in Applied Sciences and Technology,
https://doi.org/10.1007/978-981-95-6315-9_3

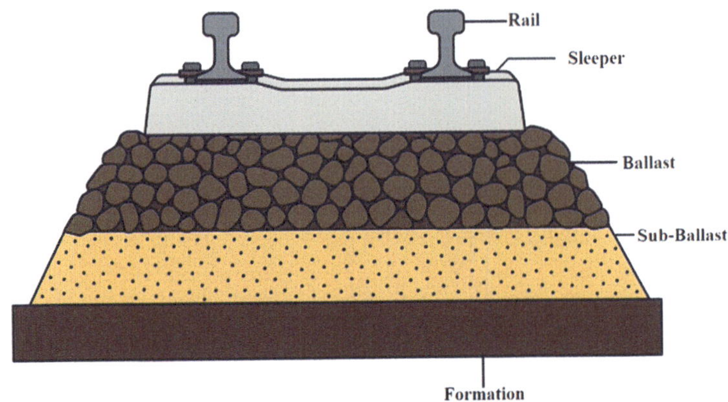

Fig. 3.1 Cross-section of railway track

per the Indian standards are 370 kN, 60 kN, and 52.5 kN in the case of rail seat test, centre top bending test and centre bottom bending tests, respectively [7].

To ensure experimental feasibility, a one-third scale model of the T-2495 sleeper (Fig. 3.2) design, widely used in Indian Railways was considered [8]. The scaled specimens were cast using 60 N/mm^2 concrete, reinforced with four high-tensile prestressing wires of 4 mm diameter. The concrete variants studied include ordinary concrete, rubcrete, steel fibre-reinforced concrete, steel fibre-reinforced rubcrete, polypropylene fibre-reinforced concrete, and polypropylene fibre-reinforced rubcrete.

The experimental programme consisted of three standard static tests: rail seat static test, centre bottom bending test, and centre top bending test, followed by a repeated drop-weight impact test. These tests were designed to emulate the mechanical actions that sleepers undergo in service and to benchmark the performance of fibre-reinforced rubcrete variants against ordinary concrete.

The experimental results were subsequently used to validate numerical models of the scaled models [9]. Simulations were performed using ANSYS Mechanical APDL for static tests and ANSYS Workbench (Explicit Dynamics) for impact analysis. Once validated, the numerical models were extended to evaluate the structural behaviour of full-scale prototypes under similar loading conditions. The integrated experimental-numerical approach enabled a rigorous comparison of material variants and provided insight into the viability of rubcrete and fibre-reinforced rubcrete for sleeper applications. In particular, these materials were evaluated in terms of load-bearing capacity, crack propagation, and energy absorption characteristics under critical loading scenarios.

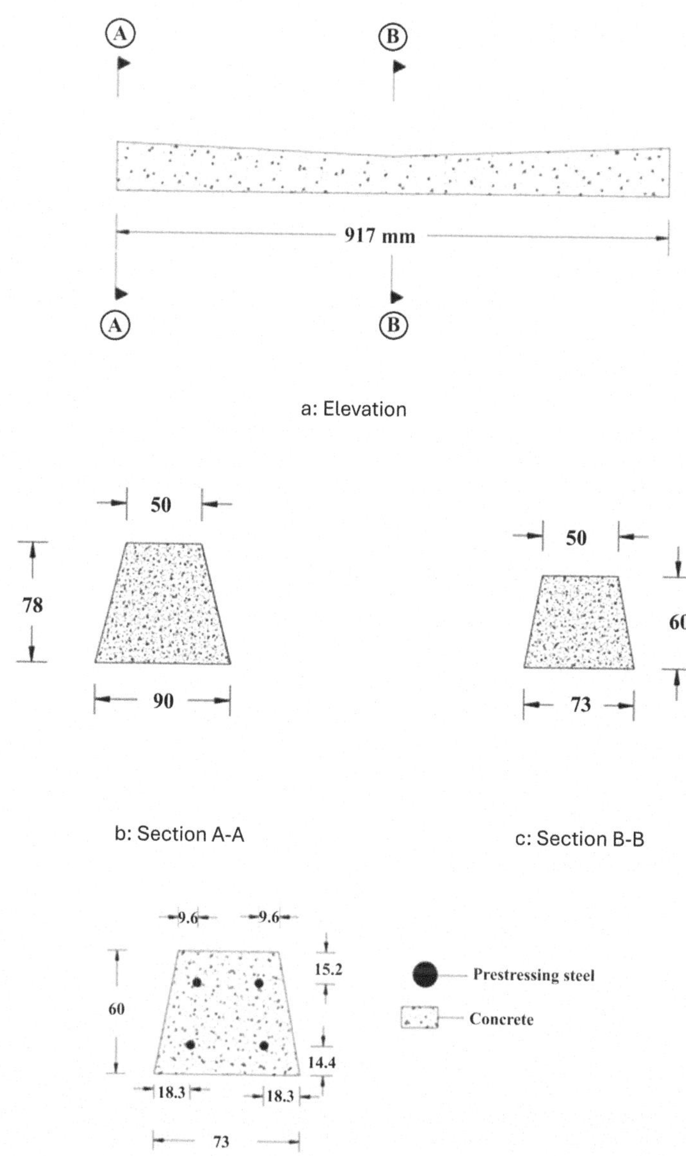

Fig. 3.2 One-third prestressed concrete sleeper model. (**a**) Elevation. (**b**) Section A-A. (**c**) Section B-B. (**d**) Location of prestressing steel (All dimensions in mm)

3.2 Experimental Investigations

In order to systematically evaluate the structural performance of different concrete variants under realistic service conditions, an experimental investigation was conducted using a one-third scaled model of the standard Indian Railway sleeper designated as T-2495. This approach was adopted to facilitate ease of handling and reduce the resource intensiveness associated with full-scale testing. Each sleeper model measured approximately one-third the size of the actual prototype but retained the critical geometric proportions and reinforcement configurations. Concrete of compressive strength 60 N/mm^2 was chosen as the matrix material to ensure high strength and stiffness (ref Chap. 2). Prestressing was achieved using four high-tension steel wires of 4 mm diameter, each possessing a minimum tensile strength of 1800 N/mm^2. These wires were inserted through precisely positioned ducts and tensioned using a hydraulic jack. Once stressed, the wires were locked in place with wedges and barrels to prevent slippage during concreting. Figure 3.3 depicts the process of production of sleepers in the laboratory [10].

Following concreting, the top surface of the moulds was covered with water-soaked gunny bags to facilitate curing. After 7 days, the prestressing wires were carefully released. The sleepers were then subjected to continued curing for an additional 21 days using moist gunny bags wrapped around the entire specimen. Before testing, the surface was given a thin coat of whitewash to enhance the visibility of crack formation. Six different material variants were considered. The designation of each specimen and the number of specimens cast per variant are listed in Table 3.1.

3.2.1 Static Testing of Sleepers

To evaluate the load-bearing capacity and failure mechanisms of the prestressed concrete sleeper models under service-like conditions, three key static tests were conducted in accordance with the guidelines of the Research Designs and Standards Organisation (RDSO):

1. **Rail Seat Static Test**
2. **Centre Bottom Bending Test**
3. **Centre Top Bending Test**

These tests were chosen to simulate critical stress states experienced by railway sleepers in actual track conditions. Each test was carried out on all six material variants of the sleeper, with three specimens tested per variant to ensure repeatability and reliability of results.

Rail Seat Static Test

The rail seat region is the most heavily loaded portion of a sleeper since it directly transfers vertical loads from the rail to the sleeper. To assess the performance of the sleeper in this region, a static load was applied vertically at one rail seat while the

a: Moulds for casting the scale-down model of the sleeper placed in prestressing bed with wires inserted

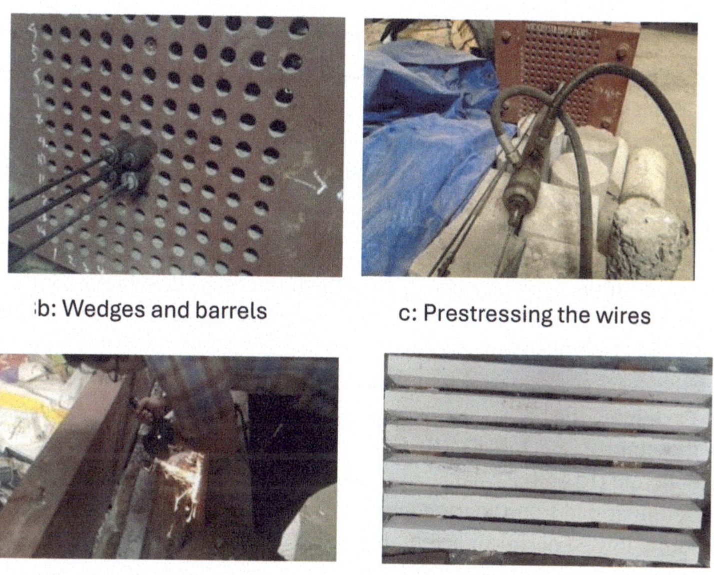

b: Wedges and barrels

c: Prestressing the wires

d: Cutting the wires

e: Sleepers after whitewash

Fig. 3.3 Stages of casting including prestressing setup, concreting, and demoulding. (**a**) Moulds for casting the scale-down model of the sleeper placed in prestressing bed with wires inserted. (**b**) Wedges and barrels. (**c**) Prestressing the wires. (**d**) Cutting the wires. (**e**) Sleepers after whitewash

opposite seat was kept unrestrained in the vertical direction. Testing was performed on a 1000 kN capacity Universal Testing Machine (UTM). The schematic diagram of the test setup is shown in Fig. 3.4. Each specimen was loaded to failure (Figs.3.5 and 3.6), and the ultimate load and failure pattern were recorded.

Table 3.1 Details of specimens

Specimen ID	Description	Number of specimens
R0	Ordinary concrete	12
R15	Rubcrete (15% crumb rubber)	12
R0PP0.2	Polypropylene fibre-reinforced concrete	12
R15PP0.2	Polypropylene fibre-reinforced rubcrete	12
R0SF0.75	Steel fibre-reinforced concrete	12
R15SF0.75	Steel fibre-reinforced rubcrete	12

Note: In specimen IDs, 'R' indicates the percentage volume of fine aggregate replaced by crumb rubber; 'SF' denotes the volume percentage of steel fibres; and 'PP' indicates the volume percentage of polypropylene fibres used in the total volume of concrete

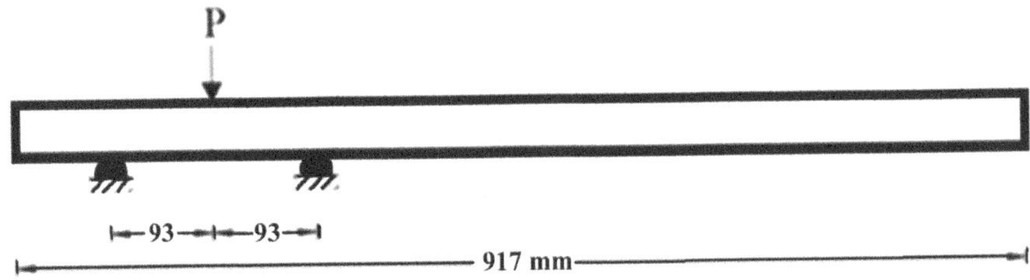

Fig. 3.4 Schematic loading arrangement of Rail Seat Static Test

The observed failure typically began with flexural cracks developing at the bottom of the sleeper, which then propagated upward, leading to concrete crushing at the top of the rail seat zone (Fig. 3.5).

Among the various material combinations, steel fibre-reinforced concrete (R0SF0.75) exhibited the highest load capacity, improving the failure load by nearly 12% over the control specimen (R0). The performance of polypropylene fibre--reinforced concrete (R0PP0.2) and steel fibre-reinforced rubcrete (R15SF0.75) was found to be comparable to ordinary concrete. Rubcrete without fibres (R15), however, showed the lowest performance in this test.

Centre Bottom Bending Test

The centre bottom bending test was conducted to simulate conditions where the bottom face of the sleeper is subjected to tension, while the top is in compression, representing typical mid-span bending under rail traffic loads. Sleepers were supported at their ends and loaded at the centre. The load was gradually increased until failure. The schematic diagram of the test setup is shown in Fig. 3.7.

As the load is increased, flexural cracks first appeared at the bottom surface in the mid-span region. With continued loading, multiple cracks formed, culminating in the crushing of concrete at the top, which marked the ultimate failure (Fig. 3.8). The ultimate loads obtained during centre bottom bending tests can be seen in Fig. 3.9.

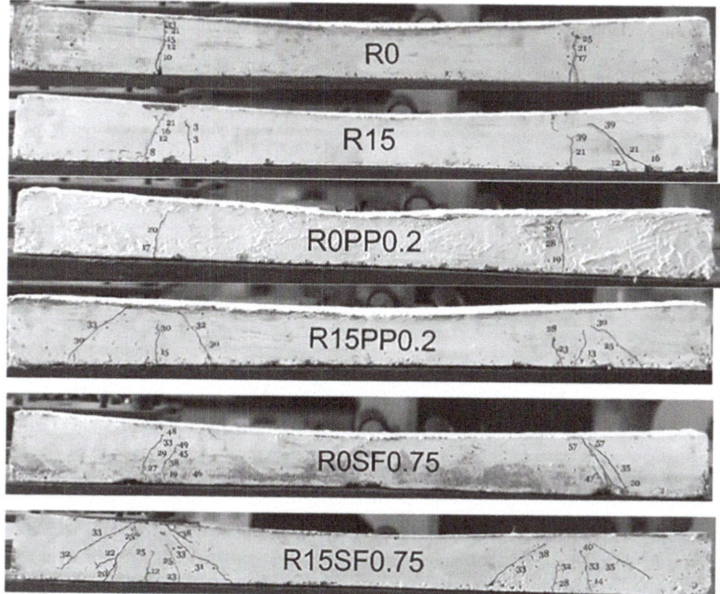

Fig. 3.5 Crack patterns observed during rail seat test

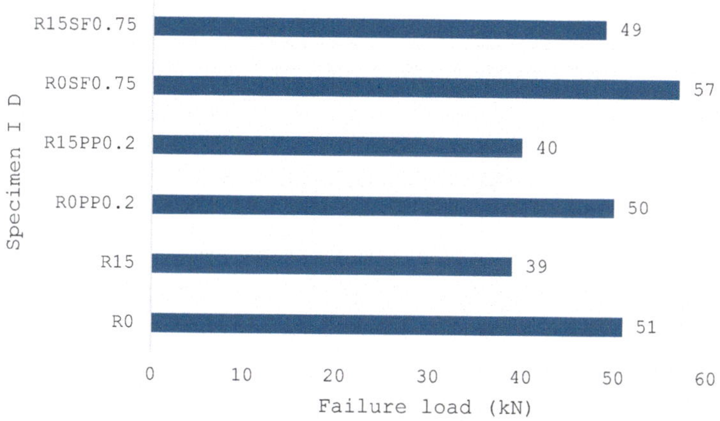

Fig. 3.6 Failure loads from rail seat static test

The steel fibre-reinforced concrete sleeper (R0SF0.75) once again demonstrated superior performance, achieving an 11.11% increase in failure load over the conventional specimen. The polypropylene fibre-reinforced concrete (R0PP0.2) and steel fibre-reinforced rubcrete (R15SF0.75) also performed comparably to the control specimen (R0). In this test also, rubcrete without fibres (R15) showed the lowest load-bearing capacity.

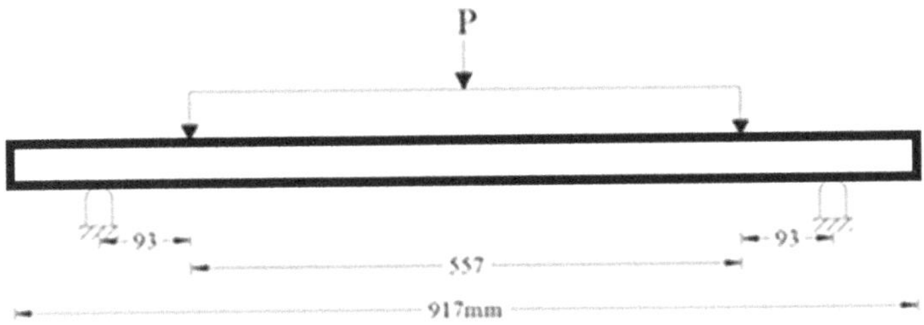

Fig. 3.7 Schematic loading arrangement of centre bottom bending test

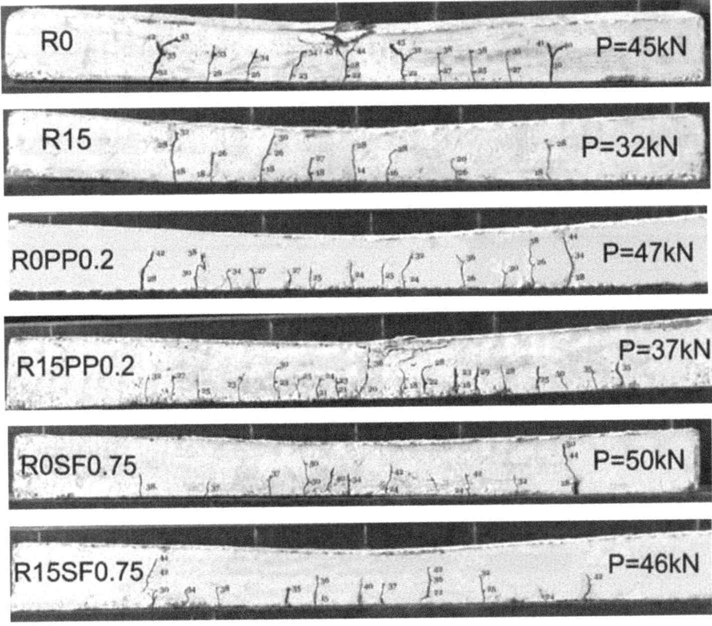

Fig. 3.8 Crack patterns after centre bottom bending

Centre Top Bending Test

The centre top bending test was designed to evaluate the behaviour of prestressed concrete sleepers when the top face is subjected to tension and the bottom face to compression: an important failure mode under certain dynamic or support-related conditions. In this test, the sleeper was simply supported and loaded at the centre of its top face until failure. The schematic loading diagram of the setup is shown in Fig. 3.10. The typical crack pattern observed in the failed specimens is illustrated in Fig. 3.11.

Flexural cracks initially formed at the top centre region of the sleeper. As the load increased, these cracks propagated downwards. The final failure was due to

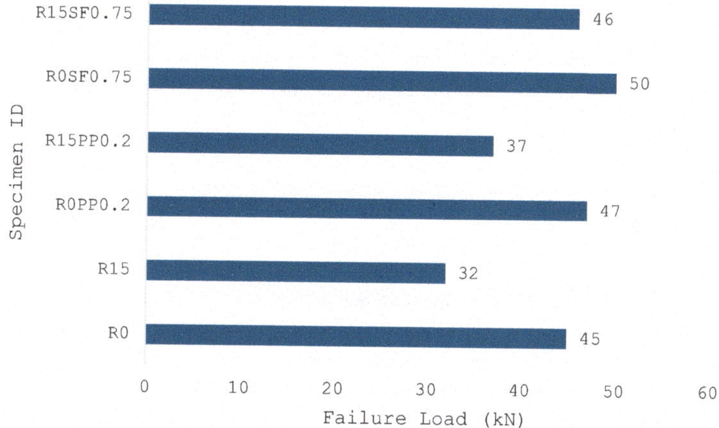

Fig. 3.9 Failure loads from centre bottom bending test

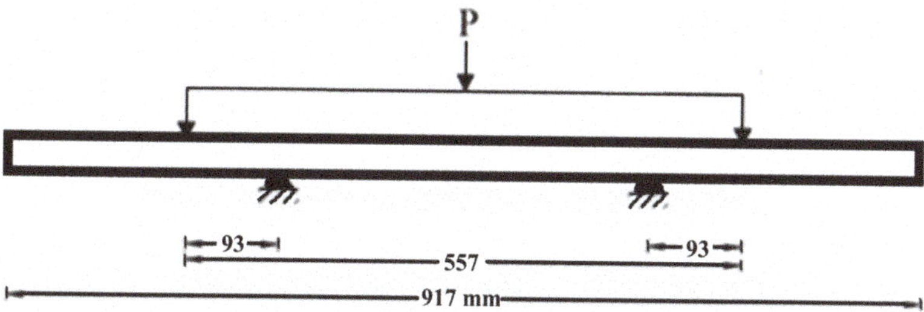

Fig. 3.10 Schematic loading arrangement of centre top bending test

crushing of the concrete at the bottom. The failure loads observed during the centre top bending tests can be noted from Fig. 3.12.

Among the tested variants, the steel fibre-reinforced concrete sleeper (R0SF0.75) exhibited the highest load-carrying capacity, with a 19.35% improvement over the ordinary concrete sleeper. The polypropylene fibre-reinforced concrete sleeper (R0PP0.2) also demonstrated enhanced performance. The rubcrete variants, especially without fibre reinforcement, lagged behind in terms of strength under this loading mode.

3.2.2 *Impact Test*

The ability of railway sleepers to withstand repeated dynamic loads is critical for their long-term performance. In real-world track systems, sleepers rest on ballast beds and are subjected to transient impacts from moving loads, especially under

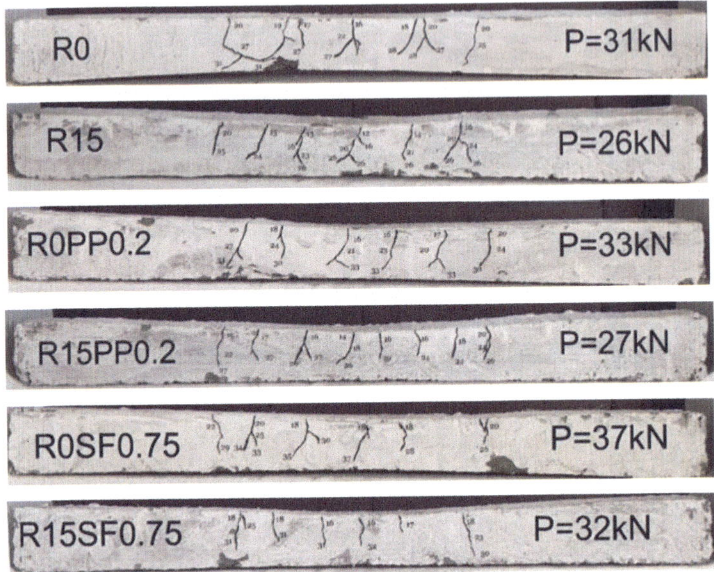

Fig. 3.11 Crack patterns after centre top bending

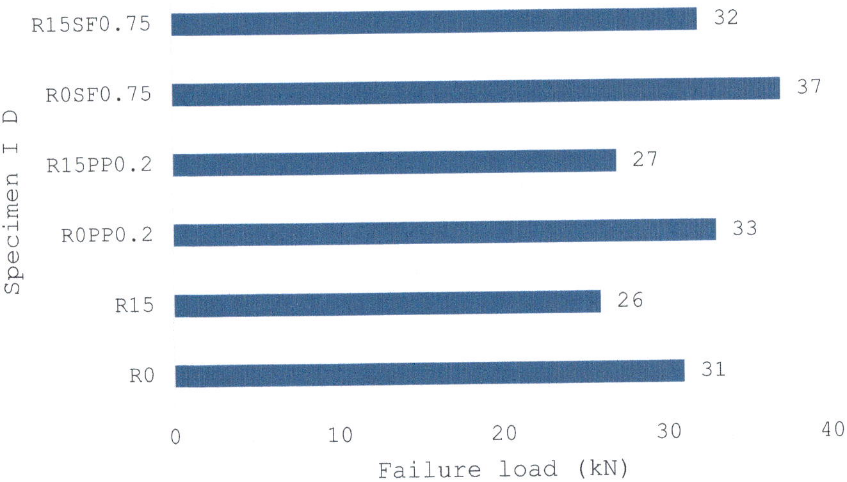

Fig. 3.12 Failure loads from centre top bending test

conditions such as wheel flat-induced pounding or sudden braking. Therefore, an impact test was carried out on all six variants of the prestressed concrete sleeper to evaluate their resistance to repeated dynamic loading.

To realistically simulate field conditions, the scale-model sleepers were placed on a ballast bed before testing. A repeated drop-weight impact test was conducted using a custom-fabricated drop-weight apparatus designed for laboratory-scale studies. The apparatus included a rigid frame, a guide rail system, and a release mechanism. A 20 kg steel impactor was suspended by a sling and released from a height of 0.4 m, with the point of impact located 100 mm off-centre from the mid-point of the sleeper, in accordance with RDSO specifications. Figure 3.13a, b depicts the impact testing of sleeper model.

The impactor was dropped repeatedly on each specimen until visible cracks were observed on the bottom surface of the sleeper. This test method enabled quantification of the sleeper's cumulative energy absorption capacity by recording the number of impacts endured before failure. The energy absorbed was calculated using Eq. (2.1):

Where the mass of the impactor was 20 kg and height of drop was 0.4 m.

The test results are presented in Fig. 3.14, and the observed crack patterns are shown in Fig. 3.15.

From the test data, it is evident that:

- The steel fibre-reinforced rubcrete sleeper (R15SF0.75) exhibited the highest energy absorption capacity, recording a cumulative energy of 3139.20 Nm, marking a 122.22% improvement over the ordinary concrete sleeper.
- The polypropylene fibre-reinforced rubcrete sleeper (R15PP0.2) also showed excellent performance, absorbing 2746.8 Nm, i.e. 94.94% higher than the control specimen.
- Rubcrete without fibres (R15) demonstrated a 66.67% increase in impact energy absorption, suggesting the inherent cushioning effect of crumb rubber.
- On the other hand, the conventional sleeper (R0) absorbed the least energy before failure, confirming its relatively brittle failure mode.

Inference from Static and Impact Tests

The experimental results from the static and impact tests provide deep insights into the structural performance of prestressed concrete sleepers made with various material combinations. The comparative evaluation across six different material variants reveals the contributions of fibres and crumb rubber to the mechanical behaviour under different loading regimes.

(a) **Static Tests**

The three static tests: rail seat static, centre bottom bending, and centre top bending were designed to reflect the primary modes of loading experienced by sleepers in service. Key observations include:

- Steel Fibre-Reinforced Concrete (R0SF0.75) consistently outperformed all other variants across all three static tests. The enhancement in ultimate load-bearing capacity of 11.76% in rail seat, 11.11% in centre bottom bending, and 19.35% in centre top bending confirms the effectiveness of steel fibres in controlling crack propagation and improving flexural strength through effective crack-bridging.
- Polypropylene Fibre-Reinforced Concrete (R0PP0.2) demonstrated marginal improvements over ordinary concrete, particularly in bending. This is attributed to the fibres' ability to delay the onset of microcracking. However, due to their

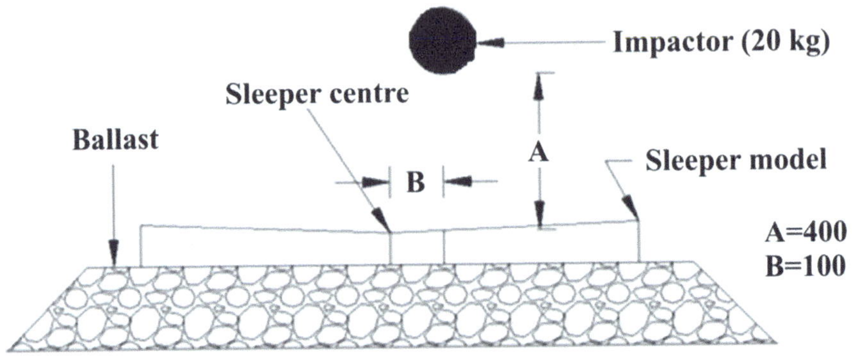

All dimensions in mm

a: Schematic diagram for impact test on sleeper

b: Laboratory test setup for impact test on sleeper

Fig. 3.13 Impact testing of sleepers. (**a**) Schematic diagram for impact test on sleeper. (**b**) Laboratory test setup for impact test on sleeper

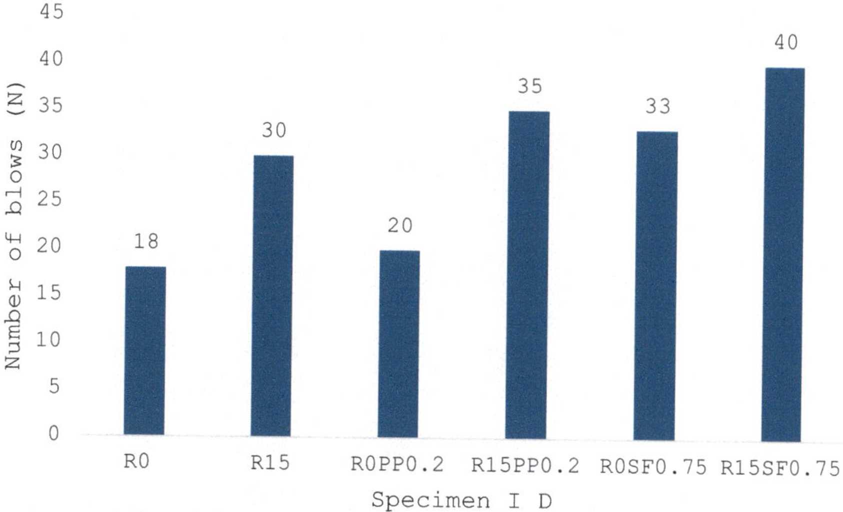

Fig. 3.14 Energy absorbed by sleeper models during impact test. (**a**) Number of blows to failure. (**b**) Cumulative energy absorbed

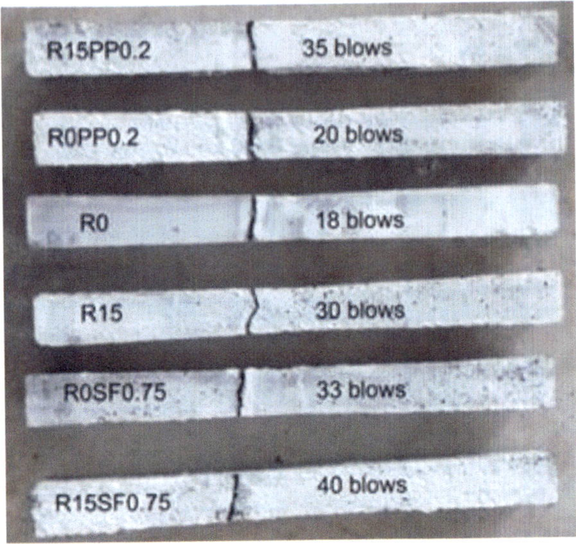

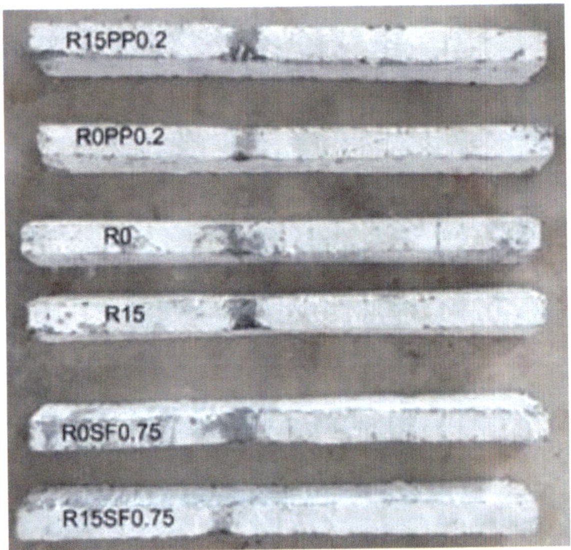

Fig. 3.15 Failure pattern of sleeper models subjected to impact test. (**a**) Splitting failure (bottom face). (**b**) Crushing due to impact (top face)

lower modulus and tensile strength compared to steel fibres, they did not significantly enhance the ultimate strength.

- Rubcrete without fibres (R15) showed considerable reduction in ultimate strength ranging from 16% to 29%, depending on the test. The inclusion of crumb rubber, while beneficial for impact resistance, led to reduced stiffness and strength in static tests.

- Steel Fibre-Reinforced Rubcrete (R15SF0.75) showed performance comparable to ordinary concrete in static conditions. The positive influence of fibres partly compensates for the strength reduction caused by crumb rubber inclusion.
- All material variants (except R15 and R15PP0.2) met the minimum strength requirements stipulated by RDSO's provisional specifications (T-39-85), especially in the simulated prototype conditions.

(b) **Impact Test**

The results from the repeated impact test clearly demonstrate the superior energy absorption capacity of fibre-reinforced rubcrete variants:

- The R15SF0.75 sleeper absorbed over 122% more energy than the ordinary concrete sleeper, a direct consequence of synergistic action between steel fibres and crumb rubber.
- R15PP0.2 also showed a 95% increase, indicating the positive contribution of rubber's damping ability and the post-crack-bridging capacity of polypropylene fibres.
- Notably, even the plain rubcrete (R15) absorbed 66.67% more energy than R0, underscoring the effectiveness of rubber particles in improving resilience under dynamic loads.
- Ordinary concrete (R0), while strong in static conditions, fractured early under impact, highlighting its brittle nature and the importance of toughness in dynamic environments.

Overall, the static and impact test results illustrate a clear trade-off between strength and energy absorption. While crumb rubber inclusion leads to reduced stiffness and strength under static loading, it significantly enhances energy dissipation under dynamic conditions. Fibre reinforcement, especially with steel fibres, successfully bridges this gap by improving both strength and toughness. These findings serve as a foundation for assessing the suitability of these composite materials for use in full-scale sleeper applications.

3.3 Numerical Simulations

The numerical studies on sleepers include the development of a numerical model to simulate the performance of prestressed concrete sleeper models, validation of numerical model with experimental results, analyse the performance of sleeper prototype using the validated numerical model.

3.3.1 Model Validation Using Scaled Sleepers

Before advancing to prototype-level simulations, it is critical to verify that the numerical model replicates experimental responses with reasonable accuracy. For this purpose, simulations were carried out on one-third scaled prestressed concrete sleepers, and the results were benchmarked against corresponding experimental data. Validation was conducted for the following tests: rail seat static test, centre bottom bending test, centre top bending test, and impact test. The key outcomes of each comparison are presented below.

Static and Bending Tests—Scaled Model

Simulations for the rail seat static test, centre top bending test, and centre bottom bending test were performed using the ANSYS Mechanical APDL module [9, 11]. Concrete was modelled using SOLID65 elements; while prestressing steel was represented with LINK180 elements. Prestressing was simulated by applying an initial strain to the tendons = 0.00398 for the scale model. Mesh generation, guided by convergence studies, yielded 6840 nodes and 5643 elements for the scale model. Details of the material properties employed in these simulations are provided in Chap. 2. The William and Warnke failure model were employed to simulate the behaviour of concrete. Cracking was assumed to initiate once the principal tensile stress surpassed the specified threshold, whereas crushing was considered to occur when the principal compressive stress exceeded the defined yield surface. The numerical simulation of the rail seat static test, centre bottom bending test, and centre top bending test was conducted using the boundary conditions and loading setup described in Fig. 3.16. The predicted ultimate loads for rail seat static test showed strong agreement with the experimental values (Table 3.2). Simulations for the centre bottom bending test captured the overall stiffness and load-carrying characteristics of the sleepers effectively. The numerical model was able to replicate the failure trends under centre top bending with high accuracy. The crack patterns obtained from the simulation (Fig. 3.17) were also in agreement with those observed in experiments.

Impact Test—Scaled Model

The impact behaviour of prestressed concrete sleepers was numerically simulated using the Explicit Dynamics module in ANSYS Workbench, which is well-suited for analysing structural responses under short-duration dynamic loads. The geometric modelling of both concrete and prestressing steel components was initially carried out in AutoCAD and subsequently imported into ANSYS Workbench's Design Modeler environment. Material properties for 60 N/mm^2 concrete, rubcrete, steel fibre-reinforced concrete, polypropylene fibre-reinforced concrete, steel fibre-reinforced rubcrete, and polypropylene fibre-reinforced rubcrete detailed in Chap. 2 were assigned to the respective elements. In the scaled model, a prestressing force of 10 kN was applied to the tendons. Mesh refinement was guided by convergence studies. The scaled model consisted of 9048 elements and 13,197 nodes, using a combination of quadrilateral and triangular elements to represent concrete, prestressing steel, and the impactor. A longitudinal contact stiffness of 50,000 N/mm was assigned between the concrete sleeper and ballast interface. To evaluate impact performance, the velocity of the impactor was gradually increased in increments of 0.1 m/s. The sleeper model was examined for failure, identified by the formation of cracks at the bottom and marked by the deletion of elements. The simulation employed the principal stress failure criterion to detect material failure under

Fig. 3.16 Loading diagram during numerical investigations on sleeper models. (**a**) Loading diagram of sleeper model subjected to rail seat static test. (**b**) Loading diagram of sleeper model subjected to centre bottom bending test. (**c**) Loading diagram of sleeper model subjected to centre top bending test

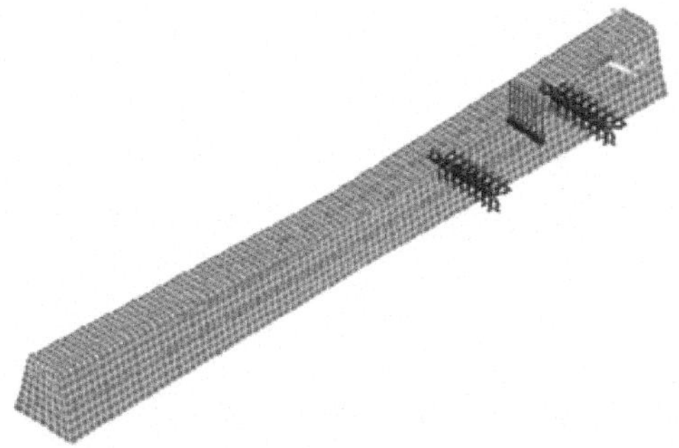

a Loading diagram of sleeper model subjected to rail seat static test.

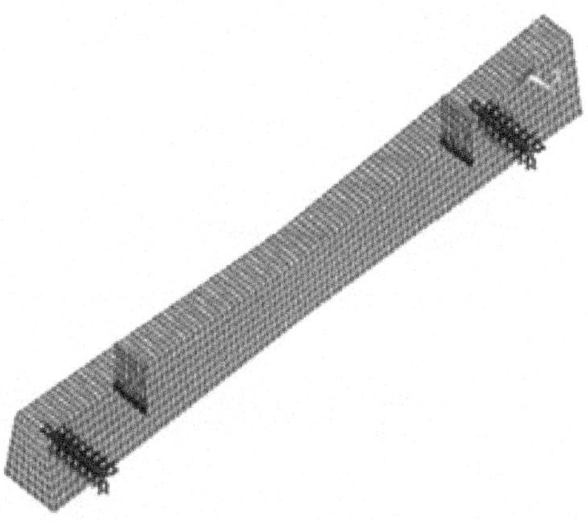

b Loading diagram of sleeper model subjected to Centre Bottom Bending Test.

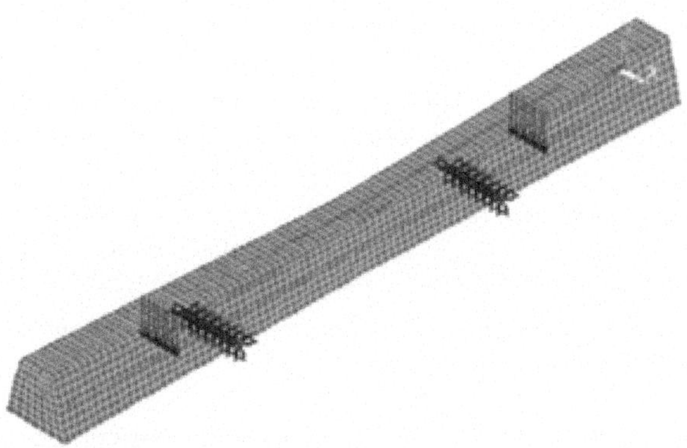

c Loading diagram of sleeper model subjected to Centre Top Bending Test.

Table 3.2 Ultimate loads observed during numerical simulation of scaled models (kN)

	Rail seat static test		Centre top bending test		Centre bottom bending test	
Specimen ID	Numerical	Experimental	Numerical	Experimental	Numerical	Experimental
R0	50	51	33	31	47	45
R15	39	39	30	26	36	32
R0PP0.2	49	50	33	33	48	47
R15PP0.2	37	40	31	27	36	37
R0SF0.75	56	57	37	37	53	50
R15SF0.75	48	49	33	32	45	46

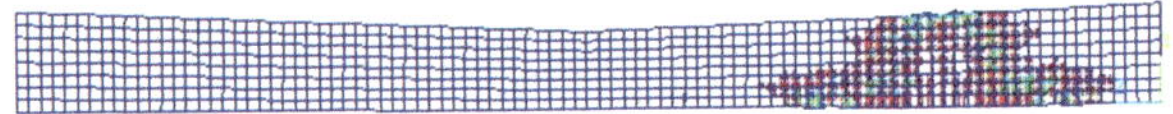

a Simulated Crack Pattern – Rail Seat Static Test (Scaled Model).

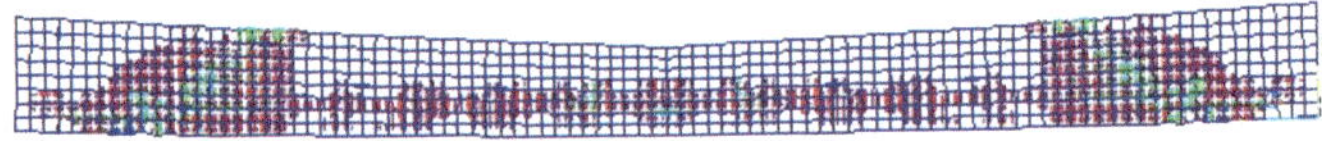

b Simulated Crack Pattern – Centre Bottom Bending Test (Scaled Model).

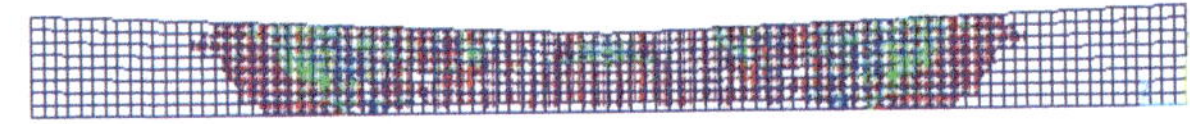

c Simulated Crack Pattern – Centre Top Bending Test (Scaled Model).

Fig. 3.17 Crack patterns of scaled sleeper models obtained from FEM. (**a**) Simulated crack pattern—rail seat static test (scaled model). (**b**) Simulated crack pattern—centre bottom bending test (scaled model). (**c**) Simulated crack pattern—centre top bending test (scaled model)

dynamic loading. The energy absorption values obtained numerically were closely matched with experimental results, as shown in Table 3.3. Loading diagram and failure initiation and crack propagation were visually comparable to those observed during repeated drop-weight tests (Fig. 3.18). The energy required to cause failure in the sleeper was subsequently calculated using Eq. (3.1). This provided a quantitative measure of the impact energy absorbed by the sleeper up to the point of failure, enabling a comparative evaluation of the performance of different material variants under dynamic loading.

$$U = 0.5\mathrm{mv}^2 \tag{3.1}$$

where U is the energy absorbed (in Nm), m is the mass of the impactor (in kg), and v is the velocity of the impactor (in m/s) at the instant when cracks are first observed in the numerical model.

Table 3.3 Comparison of impact energy for scaled model

Specimen ID	Velocity at which the failure occurred (m/s)	Impact energy (Nm)	
		Numerical	Experimental
R0	11.8	1392.4	1412.64
R15	14.8	2190.4	2354.4
R0PP0.2	12.6	1587.6	1569.6
R15PP0.2	17.3	2992.9	2746.8
R0SF0.75	16.2	2624.4	2589.84
R15SF0.75	17.7	3132.9	3139.20

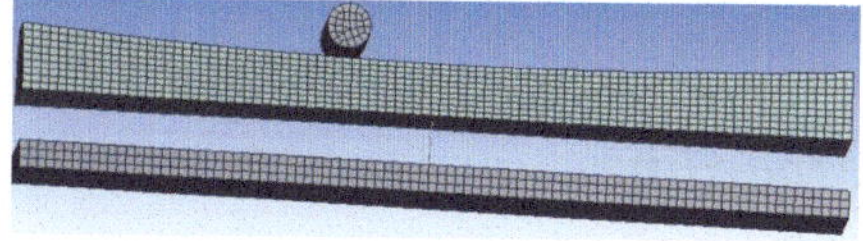

a Loading diagram of sleeper model subjected to impact loads.

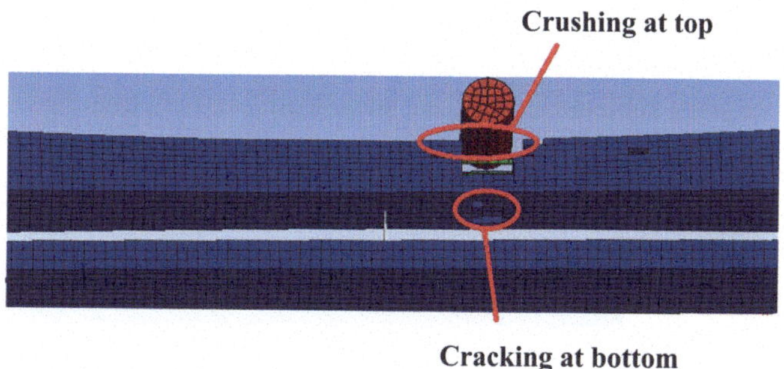

b Simulated Crack Pattern – Impact Test (Scaled Model).

Fig. 3.18 Numerical simulation of impact test on sleeper model. (**a**) Loading diagram of sleeper model subjected to impact loads. (**b**) Simulated crack pattern—impact test (scaled model)

3.3.2 Validation of Numerical Models

Validation of the finite element models was carried out by comparing the results from the numerical simulations with those obtained from the experimental investigations on the one-third scaled prestressed concrete sleepers. Four tests were considered for the validation process: rail seat static test, centre bottom bending test, centre top bending test, and impact test. Tables 3.2 and 3.3 provide a consolidated comparison of the numerical and experimental results for all tests on the scaled sleeper model.

The numerical simulations demonstrated a high degree of correlation with the experimental data. The rail seat static tests, centre bottom bending test, and centre top bending test showed good agreement, validating the load transfer mechanisms captured in the simulation. The impact test, which was simulated using the Explicit Dynamics module in ANSYS Workbench, also yielded results that were highly consistent with the physical tests. Energy absorption values from the numerical analysis matched closely with those from the laboratory impact tests, with, indicating that the selected failure criterion and material models adequately captured the dynamic response of the sleepers.

Overall, the successful validation across all four testing modalities affirms that the numerical models accurately represent the physical behaviour of the scaled sleeper models. These validated models were then employed for further studies on the full-scale prototype sleepers.

3.3.3 Prototype Modelling

With the validation of the numerical model established through the simulation of one-third scaled sleepers, the next logical step involved extending the simulations to full-scale (prototype) prestressed concrete sleepers. This section presents the methodology and results of numerical simulations carried out on prototype models of the six sleeper variants, subjected to rail seat static load, centre bottom bending, centre top bending, and impact loading.

The finite element modelling for the prototype sleepers followed a similar approach to that adopted for the scaled models, with appropriate refinements in geometry, mesh density, boundary conditions, and prestressing force. The solid elements used for concrete (SOLID65 in ANSYS Mechanical APDL and its equivalent in Workbench) and link elements for prestressing wires were maintained. The prestressing force applied to the wires in the prototype model was 27 kN. Prestressing was simulated through an initial strain input of 0.0064 in the high-tension wires, as in the earlier prototype simulations.

Meshing was carried out based on convergence studies to ensure accuracy and computational efficiency. The prototype model comprised 54,579 nodes and 51,308 elements for static simulations in ANSYS Mechanical APDL, and 59,256 nodes and 20,214 elements for dynamic simulations in ANSYS Workbench. Material properties used in the model were consistent with those experimentally obtained and previously presented in Chap. 2.

The loading patterns and boundary conditions for each type of test (rail seat static, centre bottom bending, centre top bending, and impact) were applied in accordance with RDSO specifications. For the impact test, the explicit dynamics module in ANSYS Workbench was employed to simulate high strain-rate conditions. Failure was defined by the initiation and propagation of cracks, identified by element erosion in the impacted zones.

The results from each simulation including ultimate loads, crack patterns, and energy absorption were compared against codal provisions (T-39-85) (Research Designs and Standards Organisations (2011)). The findings provide critical insights into the structural performance of each material variant, thereby supporting the

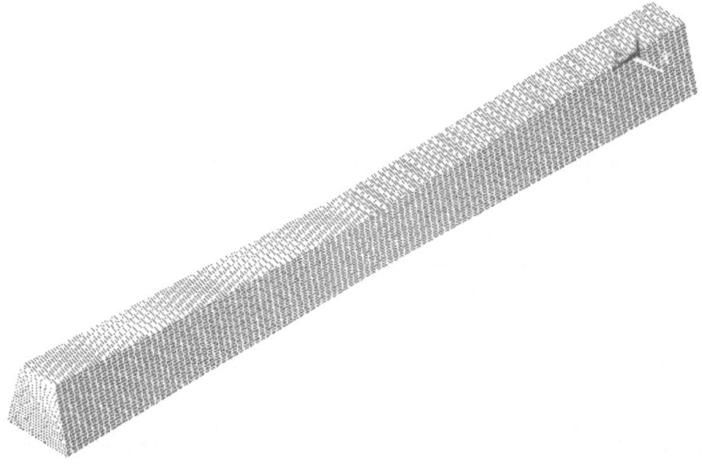

a Numerical model of prototype sleeper – mesh configuration.

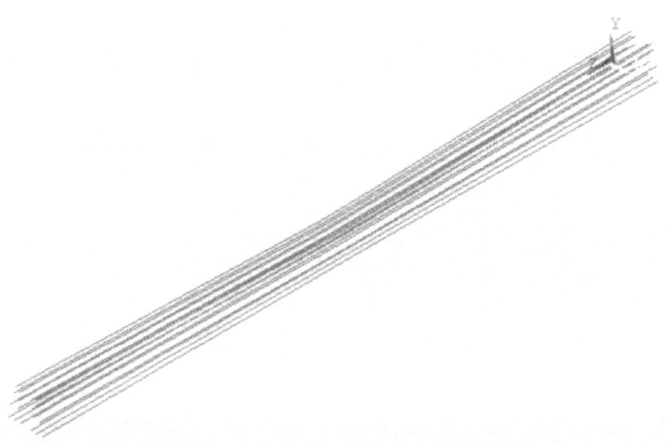

b Prestressing wire layout – prototype sleeper.

Fig. 3.19 Numerical modelling of sleeper prototypes for static and bending tests. (**a**) Numerical model of prototype sleeper—mesh configuration. (**b**) Prestressing wire layout—prototype sleeper

feasibility assessment of their use in actual railway sleeper applications. Figure 3.19 depicts the numerical model of prototype of prestressed concrete sleepers.

To evaluate the structural integrity and load-carrying capacity of the prototype prestressed concrete sleepers under real-scale conditions, numerical simulations of the rail seat static test, centre bottom bending tests, and centre top bending tests were performed in ANSYS Mechanical APDL module. The validated modelling framework from the scaled sleeper analysis was adapted for this purpose, incorporating the full dimensions, boundary conditions, and prestressing force as applicable

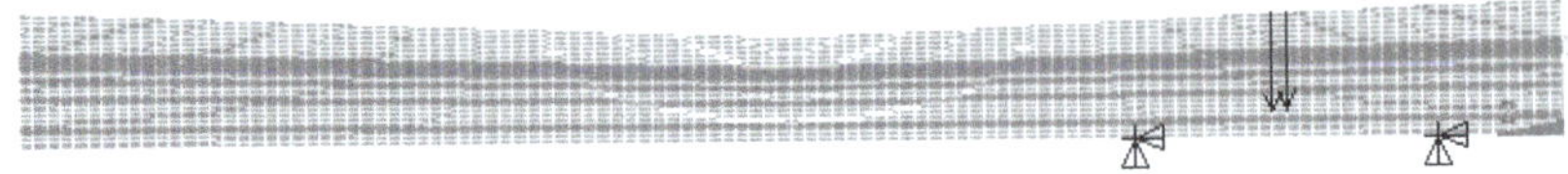

a Load application and boundary condition setup for Rail Seat Static Test.

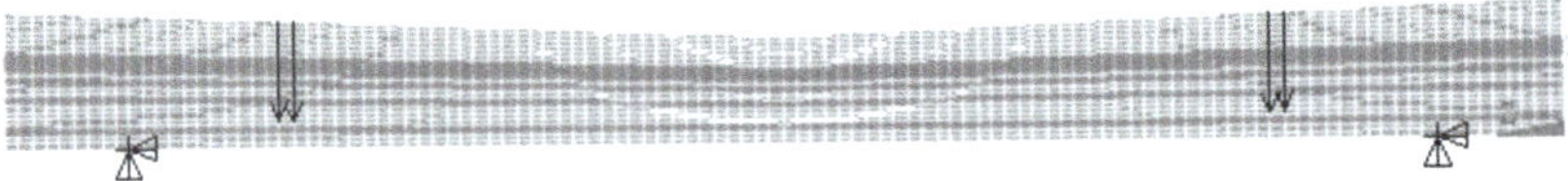

b Load and boundary conditions for Centre Bottom Bending Test.

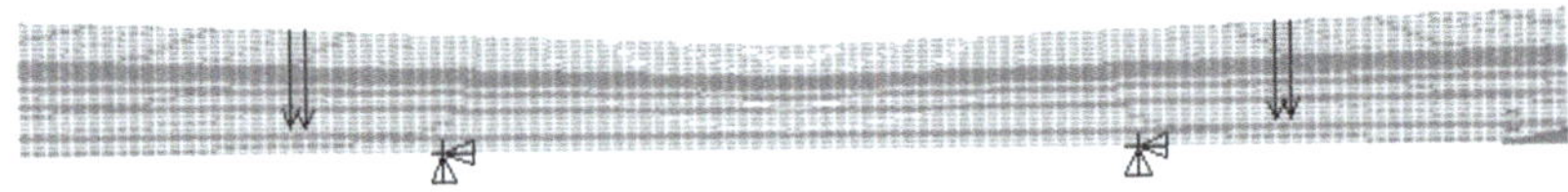

c Load and support conditions for Centre Top Bending Test.

Fig. 3.20 Loading arrangement for static and bending tests of sleeper prototypes. (**a**) Load application and boundary condition setup for rail seat static test. (**b**) Load and boundary conditions for centre bottom bending test. (**c**) Load and support conditions for centre top bending test

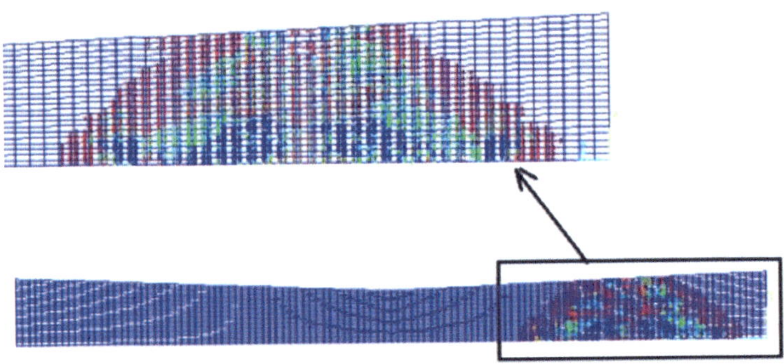

Fig. 3.21 Crack propagation pattern—rail seat static test (prototype sleeper)

to the prototype. The sleeper model was supported and loaded in accordance with the provisions of RDSO's T-39-85 specifications. The load and support conditions used in the simulation of static and bending tests are illustrated in Fig. 3.20.

The results of the simulation, including ultimate load capacity and failure patterns, are shown in Figs. 3.21, 3.22, and 3.23 and Table 3.4.

Table 3.4 presents a comparison between the simulated ultimate loads and the codal requirement as per T-39-85. In the case of rail seat static test, it is observed that the ordinary concrete (R0), polypropylene fibre-reinforced concrete (R0PP0.2),

Fig. 3.22 Crack propagation pattern—centre bottom bending test (prototype sleeper)

Fig. 3.23 Crack propagation pattern—centre top bending test (prototype sleeper)

Table 3.4 Ultimate loads observed during numerical simulation of prototypes (kN)

Specimen ID	Rail seat static test		Centre top bending test		Centre bottom bending test	
	Required	Observed	Required	Observed	Required	Observed
R0	370	480	60	69	52.5	190
R15		360		63		170
R0PP0.2		470		70		190
R15PP0.2		350		67		160
R0SF0.75		550		81		210
R15SF0.75		490		74		190

steel fibre-reinforced concrete (R0SF0.75), and steel fibre-reinforced rubcrete (R15SF0.75) variants exceeded the minimum required capacity, indicating their suitability for field application. In contrast, the rubcrete without fibres (R15) and polypropylene fibre-reinforced rubcrete (R15PP0.2) variants fell short of the requirement. In rail seat static test, the initial cracks were observed at the bottom of the rail seat. The crack propagation was then observed near the rail seat region. This was in agreement with that of experimental observations.

In centre bottom bending test simulation, steel fibre-reinforced concrete (R0SF0.75) demonstrated the highest load capacity with a 10.5% improvement over the ordinary concrete sleeper (R0). Polypropylene fibre-reinforced and rubcrete-based sleepers, while meeting the code provisions, exhibited lower performance in comparison to fibre-reinforced concrete. In centre bottom bending tests, the tensile stress developed at the bottom fibre due to bending action led to crack formation and eventual failure. The nature of crack propagation was observed to be similar to that seen in scaled model experiments and numerical simulations, thereby strengthening the reliability of the model.

In centre top bending tests, the highest load-carrying capacity was recorded for the steel fibre-reinforced concrete sleeper (R0SF0.75), which exhibited a 17.4% increase over ordinary concrete. Steel fibre-reinforced rubcrete (R15SF0.75) also demonstrated significant improvement with a 7.2% enhancement. The rubcrete (R15) and polypropylene fibre-reinforced rubcrete (R15PP0.2) sleepers, while compliant, showed reduced capacity due to the inherent loss in compressive strength associated with rubber inclusion. The initial crack development in centre top bending tests was observed at the top of the sleepers. Then the cracks propagated towards the bottom.

To evaluate the response of full-scale sleepers under dynamic loading conditions, the impact behaviour of the prototype prestressed concrete sleepers was simulated using the *Explicit Dynamics* module of ANSYS Workbench. This test captures the effects of repeated, localized impacts which sleepers may experience due to ballast voids, dropped loads, or track disturbances.

The sleeper geometry and prestressing wire layout were first modelled in AutoCAD and then imported into the ANSYS Design Modeler interface. Concrete and steel were modelled with material properties consistent with those described in Chap. 2. Prestressing was simulated by applying a 27 kN force at both ends of each tendon, corresponding to the prestress level in the prototype.

The sleepers were supported over an idealised ballast layer. A longitudinal stiffness of 50,000 N/mm was used to represent the ballast-sleeper interface. The impactor, having a mass of 20 kg, was modelled as a rigid body. It was released from varying velocities to determine the critical energy at which visible cracking or material failure (identified through element deletion) occurred. The failure criterion adopted was the *maximum principal stress* failure model.

The numerical model and prestressing layout are shown in Figs. 3.24 and 3.25, and the crack pattern observed at failure is depicted in Fig. 3.26. The velocity at

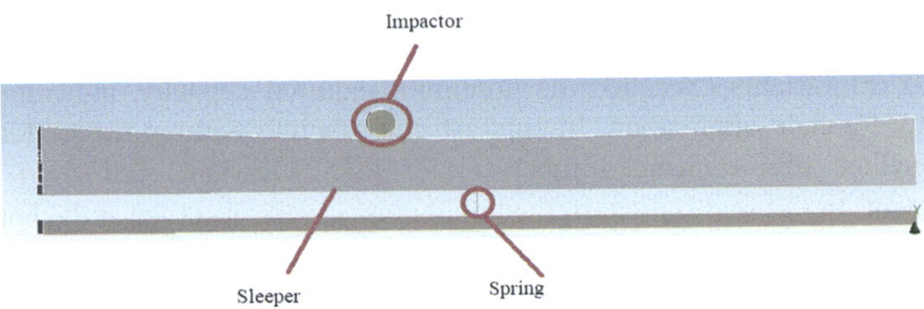

Fig. 3.24 Numerical model of prototype sleeper for impact simulation

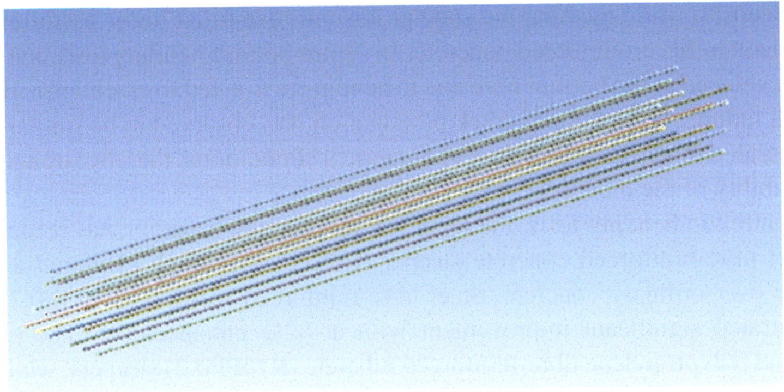

Fig. 3.25 Prestressing layout and meshing for prototype sleeper

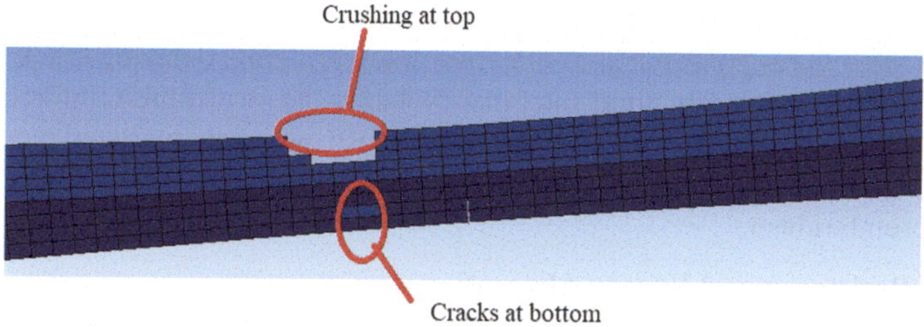

Fig. 3.26 Crack pattern—impact simulation on prototype sleeper

Table 3.5 Energy absorbed—impact test on prototype sleepers

Specimen ID	Velocity at failure (m/s)	Energy absorbed (kNm)	Percentage increase
R0	9.3	8.649	–
R15	11.7	13.689	58.3
R0PP0.2	9.5	9.025	4.4
R15PP0.2	11.8	13.924	61
R0SF0.75	12.3	15.129	75
R15SF0.75	13.2	17.424	101.5

which the first visible crack occurred was recorded, and the corresponding energy absorbed was calculated using the kinetic energy expression used in Eq. (3.1).

Table 3.5 presents the numerical simulation results of the impact energy absorbed by each sleeper variant. The findings indicate that all fibre-reinforced variants outperformed the ordinary concrete sleeper (R0) in terms of energy absorption. Notably, the steel fibre-reinforced rubcrete sleeper (R15SF0.75) demonstrated the highest energy absorption, with an increase of over 100% compared to R0. The results were found to be similar with those obtained from the scaled model experiments and confirm the superior impact resistance imparted by steel fibres, especially when used in conjunction with pre-treated crumb rubber.

3.3.4 Discussion

The comprehensive evaluation of six material variants through experimental and numerical investigations on prestressed concrete sleepers offers critical insights into their mechanical and dynamic performance. This section summarises the findings from static and impact tests, highlighting comparative trends, failure mechanisms, and the effects of adding fibres and rubber in concrete.

Performance Under Static Loading

Across all three static tests—rail seat loading, centre bottom bending, and centre top bending—the steel fibre-reinforced concrete (R0SF0.75) consistently outperformed the control specimen (R0) in terms of ultimate load capacity. The presence of crimped steel fibres contributed to effective crack-bridging, delaying crack propagation and enhancing structural integrity. Notably, the steel fibre-reinforced rubcrete (R15SF0.75) also performed comparably to R0, with only marginal reductions in strength in some configurations, despite the inherent loss of strength due to addition of crumb rubber.

Impact Resistance and Energy Absorption

The impact performance of sleeper variants revealed that the inclusion of fibres significantly enhanced energy absorption capacity. Steel fibre-reinforced rubcrete (R15SF0.75) demonstrated the highest energy absorption in both experimental and numerical tests more than doubling the capacity of ordinary concrete sleepers. This confirms the combined effect of rubber and steel fibres in enhancing toughness and post-crack energy dissipation.

Advantage of Fibre-Reinforced Rubcrete Sleepers

The integration of fibres into rubberised concrete has demonstrated considerable potential in advancing sustainable and durable sleeper technologies. Fibre-reinforced rubcrete sleepers offer a dual advantage in environmental and mechanical aspects. The incorporation of crumb rubber reduces the demand for natural aggregates and addresses the issue of tyre waste management. At the same time, the addition of steel or polypropylene fibres mitigates the strength loss typically associated with rubber inclusion, restoring and often enhancing load-bearing and impact-resistant performance.

Steel fibre-reinforced rubcrete sleepers (R15SF0.75), in particular, showed superior energy absorption, making them ideal for dynamic track environments with high-impact loads. These sleepers are expected to exhibit greater fatigue life and reduced maintenance needs due to their improved crack resistance and energy dissipation characteristics. Moreover, the use of rubber and fibres can contribute to sleepers with enhanced toughness, making handling and installation more efficient.

3.4 Benefit to Cost Ratio of Sleepers

Rubcrete sleepers without fibres and polypropylene fibre-reinforced rubcrete sleepers were found to be inadequate in meeting the performance standards required for prestressed concrete sleepers. However, a notable enhancement in energy absorption capacity was observed in the other four material variants. Consequently, the energy absorbed to cost ratio was evaluated for the following scaled-down sleeper models: ordinary concrete, polypropylene fibre-reinforced concrete, steel fibre-reinforced concrete, and steel fibre-reinforced rubcrete. These values are presented in Table 3.6 to identify the most suitable sleeper among those examined. The steel fibre-reinforced rubcrete sleeper exhibited the highest performance, demonstrating a 101% improvement in energy absorbed to cost ratio compared to the ordinary concrete sleeper.

Table 3.6 Energy absorbed to cost ratio for scaled prestressed concrete sleepers

Specimen ID	Cost per sleeper model (₹)	Energy absorbed (Nm)	Energy absorbed to cost ratio (Nm/₹)
R0	323.16	1412.64	4.37
PP0.2	325.36	1569.60	4.82
SF0.75	353.58	2589.84	7.32
R15SF0.75	357.36	3139.20	8.78

The findings from these investigations establish steel fibre-reinforced rubcrete sleepers as the most effective among the material variants evaluated for prestressed concrete sleeper applications. With an average sleeper density of approximately 1663 units per kilometre in Indian Railways, the adoption of this material can lead to substantial environmental and material benefits. Specifically, it enables a reduction of nearly 14 tonnes of fine aggregate and the utilization of about 4 tonnes of crumb rubber per kilometre of track laid. This approach not only promotes sustainable construction but also offers a practical solution to the growing issue of waste tyre disposal. Thus, the use of steel fibre-reinforced rubcrete sleepers represents a novel and impactful initiative—contributing to the conservation of natural resources while transforming an environmental liability into a valuable construction material.

3.5 Summary

Experimental and numerical investigations were carried out on both scaled and prototype models of the T-2495 sleeper, made using six different concrete variants: ordinary concrete, rubcrete, fibre-reinforced concrete, and fibre-reinforced rubcrete. The findings from the rail seat static test, centre top bending test, centre bottom bending test, and impact test led to the following key conclusions:

- The steel fibre-reinforced rubcrete sleeper demonstrated a 101–122% increase in energy absorption capacity, equivalent to 2.0–2.22 times when compared to the conventional prestressed concrete sleeper.
- Test results from the rail seat static, centre top bending, and centre bottom bending tests confirmed that sleepers made of polypropylene fibre-reinforced concrete, steel fibre-reinforced concrete, and steel fibre-reinforced rubcrete met the minimum performance standards specified in T-39-85.
- Among all six material variants evaluated for suitability in prestressed concrete sleeper applications, steel fibre-reinforced rubcrete exhibited the highest energy absorption capacity. Its performance under both static and bending loads was found to be satisfactory.
- Replacing ordinary concrete sleepers with steel fibre-reinforced rubcrete sleepers can result in a reduction of approximately 14 tonnes of fine aggregate and the utilization of 4 tonnes of crumb rubber per kilometre of railway track constructed.

- This approach represents a novel and sustainable solution, offering both resource conservation and environmental mitigation by repurposing discarded tyre rubber into structural railway components.

For further details on the application of fibre-reinforced rubcrete on railway sleepers, the readers could also refer to the journals published by the authors [8, 11–13].

References

1. É.A. Silva, D. Pokropski, R. You, S. Kaewunruen, Comparison of structural design methods for railway composites and plastic sleepers and bearers. Aust. J. Struct. Eng. **7982**, 1–17 (2017). https://doi.org/10.1080/13287982.2017.1382045
2. S. Kaewunruen, A.M. Remennikov, Effect of a large asymmetrical wheel burden on flexural response and failure of railway concrete sleepers in track systems. Eng. Fail. Anal. **15**, 1065–1075 (2008). https://doi.org/10.1016/j.engfailanal.2007.11.013
3. J.A. Zakeri, J. Sadeghi, Field investigation on load distribution and deflections of railway track sleepers field investigation on load distribution and deflections of railway track sleepers. J. Mech. Sci. Technol. **21**, 1948–1956 (2007). https://doi.org/10.1007/BF03177452
4. S. Iwnicki, *Handbook of Railway Vehicle Dynamics* (CRC Press, Boca Raton, 2019)
5. J. Taherinezhad, M. Sofi, P.A. Mendis, T. Ngo, A review of behaviour of prestressed concrete sleepers. Electron. J. Struct. Eng. **13**, 1–16 (2013). https://doi.org/10.56748/ejse.131571
6. J. Zakeri, F.H. Rezvani, Failures of railway concrete sleepers during service life. Int. J. Constr. Eng. Manag. **1**, 1–5 (2012). https://doi.org/10.5923/j.ijcem.20120101.01
7. Research Designs and Standards Organisations, Indian railway standard specification for pre-tensioned prestressed concrete sleepers for broad gauge and metre gauge (T-39-85). New Delhi, India (2011)
8. A. Raj, P. Nagarajan, A.P. Shashikala, Investigations on fiber-reinforced rubcrete for railway sleepers. ACI Struct. J. **117**, 109–120 (2020). https://doi.org/10.14359/51724679
9. A. Raj, P. Nagarajan, S.A. Pallikkara, C. Ngamkhanong, Performance characterization of fiber-reinforced rubcrete sleepers. J. Struct. Des. Constr. Pract. **30** (2025). https://doi.org/10.1061/JSDCCC.SCENG-1743
10. A. Raj, P. Nagarajan, A.P. Shashikala, *Application of Fibre Reinforced Rubcrete in Structures Subjected to Impact Load* (National Institute of Technology Calicut, 2021)
11. A. Raj, P. Nagarajan, A.P. Shashikala, Failure prediction of impact behaviour of self-compacted rubcrete sleepers. Mater. Des. Process. Commun., 1–8 (2020). https://doi.org/10.1002/mdp2.174
12. A. Raj, P. Nagarajan, A.P. Shashikala, Failure prediction of impact behaviour of self-compacted rubcrete sleepers. Mater. Des. Process. Commun. **3**, e174 (2021). https://doi.org/10.1002/mdp2.174
13. A. Raj, P. Nagarajan, A.P. Shashikala, A review on the development of new materials for construction of prestressed concrete railway sleepers. IOP Conf. Ser. Mater. Sci. Eng. **330**, 012129 (2018). https://doi.org/10.1088/1757-899X/330/1/012129

Chapter 4
Fencing Posts

4.1 Introduction

Fencing posts serve as critical structural components in securing boundaries across various infrastructural and agricultural settings. Their primary function lies in supporting fencing wires and resisting both static and dynamic forces that may arise from environmental exposure, accidental impacts, or imposed service loads. Particularly in rural and semi-urban landscapes, concrete fencing posts are preferred for their durability, low maintenance, and resistance to decay and pest attack.

Among the different types of fencing posts, *line posts* are the most frequently installed along fence lines to act as intermediate posts. Despite the wide usage of cement concrete fencing posts, limitations in impact resistance, vulnerability to cracking, and environmental concerns surrounding aggregate consumption necessitate the exploration of alternative material systems. The integration of recycled crumb rubber and reinforcing fibres in concrete especially in the form of rubcrete and its fibre-reinforced variants offers a potential route toward enhancing energy absorption capacity while addressing sustainability concerns.

This chapter presents a comprehensive investigation into the behaviour of fencing posts made with ordinary concrete, rubcrete, fibre-reinforced concrete, and fibre-reinforced rubcrete. Both experimental and numerical methods are employed to evaluate the structural performance of the fencing posts under static and impact loading conditions. Not only the mechanical adequacy of these alternative mixes but also their cost-effectiveness and material efficiency were assessed, thereby contributing to the development of more sustainable fencing solutions.

A. Raj et al., *Rubcrete and Its Applications in Structures Subjected to Impact Loading*, SpringerBriefs in Applied Sciences and Technology,
https://doi.org/10.1007/978-981-95-6315-9_4

4.2 Experimental Investigations

This section presents the experimental and numerical investigations carried out on reinforced concrete fencing posts fabricated using six different material variants: ordinary concrete, rubcrete, polypropylene fibre-reinforced concrete, polypropylene fibre-reinforced rubcrete, steel fibre-reinforced concrete, and steel fibre-reinforced rubcrete. The primary objective was to understand their behaviour under static tensile loads: representative of service conditions and impact loads, simulating accidental or dynamic forces commonly encountered in the field.

The experimental programme focused on *line posts*, which form the majority of fencing elements in typical installations [1]. As per IS: 4996-1984, fencing posts should be manufactured with concrete of minimum compressive strength 20 N/mm^2. The dimensions and reinforcement details of the fencing posts used conform to those prescribed for standard line posts [2]. Figure 4.1 presents the schematic and cross-section of the fencing post specimens (Table 4.1).

The experimental programme comprised static and impact tests conducted on fencing posts to evaluate their structural performance under service and accidental loading conditions.

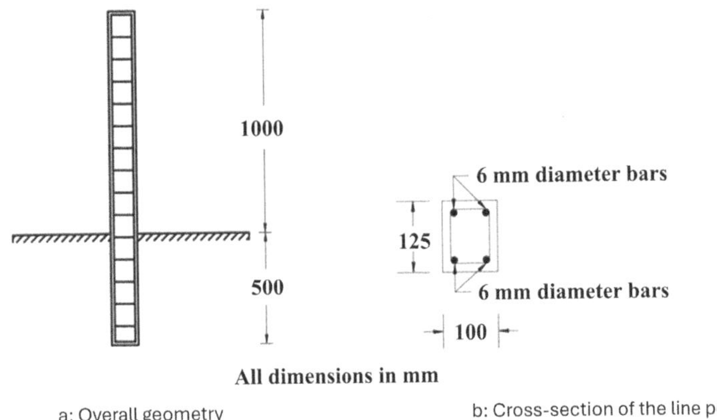

Fig. 4.1 Details of line fencing posts. (**a**) Overall geometry. (**b**) Cross-section of the line post

Table 4.1 Number of specimens

Specimen I D	Specimens cast
R0	6
R15	6
R0PP0.2	6
R15PP0.2	6
R0SF0.75	6
R15SF0.75	6

4.2.1 Static Test on Line Posts

The static test was conducted to simulate the tensile load experienced by the fencing posts when wires are tensioned between posts. The loading was applied using a 3000 kN capacity compression cum bending testing machine. Each post was placed horizontally with appropriate supports and loaded vertically as shown in Figs.4.2 and 4.3 to simulate the action of lateral pull from the wires.

The load at which the first visible crack appeared was recorded for each specimen. The results are summarised in Table 4.2.

The load-carrying capacities of all fencing post variants exceeded the minimum static load requirement of 700 N specified by the standard, confirming their adequacy under service loading.

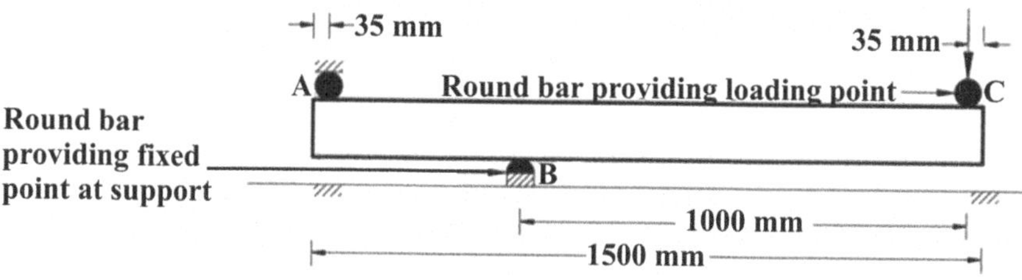

Fig. 4.2 Schematic diagram of static test on fencing post

Fig. 4.3 Test setup for static load on fencing post

Table 4.2 First crack load during static test

Specimen ID	Load (N)
R0	1200
R15	1000
R0PP0.2	1200
R15PP0.2	1000
R0SF0.75	1300
R15SF0.75	1200

Fig. 4.4 Failure patterns on fencing posts during static test (top face)

Figure 4.4 shows the typical failure patterns observed during the test. Tension cracks were primarily observed at the top face of point B (Fig. 4.2) in all the variants, with crack widths varying depending on the type of reinforcement and material.

4.2.2 Impact Test on Line Posts

To evaluate the performance of fencing posts under accidental and sudden loading conditions, impact tests were conducted following the guidelines of IS: 4996-1984. The test setup was designed to replicate the dynamic conditions that a fencing post might experience in the field, such as impact from stray animals, falling objects, or misaligned machinery.

An impactor of mass 20 kg was repeatedly dropped from a height of 115 mm onto the specimen until visible cracking was observed at the bottom face of the fencing post. This approach allowed quantification of the energy absorption capacity of each concrete variant through the number of blows sustained.

Figures 4.5 and 4.6 show the schematic representation and actual setup used for the impact tests.

The test results including number of blows endured, total energy absorbed (Eq. 2.1), and percentage improvement compared to ordinary concrete are summarised in Table 4.3.

Significant improvements in impact resistance were observed in the rubcrete and fibre-reinforced variants, particularly for steel fibre-reinforced rubcrete, which showed nearly three times the energy absorption capacity of ordinary concrete.

The failure modes observed during the impact testing are shown in Fig. 4.7. The top region of the fencing posts displayed crushing due to direct contact with the impactor, while tension cracks developed near the bottom face due to flexural stresses.

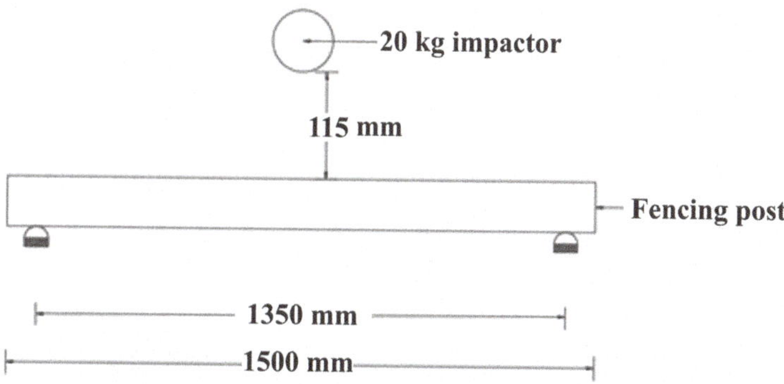

Fig. 4.5 Schematic diagram of impact test on fencing posts

Fig. 4.6 Test setup for impact test on fencing posts

Table 4.3 Energy absorbed during impact test

Specimen ID	Number of blows	Energy Absorbed (Nm)	Percentage increase (%)
R0	3	67.7	0
R15	4	90.3	33.33
R0PP0.2	4	90.3	33.33
R15PP0.2	5	112.8	66.67
R0SF0.75	6	135.4	100.00
R15SF0.75	8	180.5	166.67

Fig. 4.7 Failure patterns on fencing posts during impact test (bottom face)

4.3 Numerical Investigations

Numerical investigations were conducted to simulate and validate the structural behaviour of fencing posts subjected to static and impact loads. These simulations complemented the experimental studies by providing insight into internal stress distributions, crack development mechanisms, and energy absorption characteristics. Two separate modules within the ANSYS software suite were utilised:

- ANSYS Mechanical APDL for static load simulations.
- ANSYS Workbench (Explicit Dynamics) for impact load simulations.

This dual approach ensured that both quasi-static and dynamic responses could be accurately captured and compared against the experimental observations.

4.3.1 Numerical Investigations of Static Test on Fencing Post

The numerical simulation of the static test on fencing posts was conducted with the objective of replicating the crack patterns and first crack loads observed in experimental studies. The simulation followed a rigorous modelling process involving careful material definition, meshing, boundary condition assignment, and load application.

Element Types and Modelling Approach
Concrete was modelled using SOLID65 elements, suitable for nonlinear concrete behaviour with cracking and crushing capabilities. Reinforcement was modelled using LINK180 elements. These elements were embedded using node connectivity to ensure full bond behaviour between steel and concrete, replicating the internal mesh framework of the post. The mesh sensitivity analysis revealed that a mesh size of 15 mm yielded reliable results without excessive computational demand. The final mesh consisted of 8080 nodes and 6752 elements. One end of the post was fully fixed to replicate ground-level embedment. Figure 4.8 shows the meshed model and reinforcement layout. Figure 4.9 shows the boundary conditions and loading arrangements.

Fig. 4.8 Numerical model of fencing post. (**a**) Modelling of reinforcement. (**b**) Modelling of concrete and reinforcement

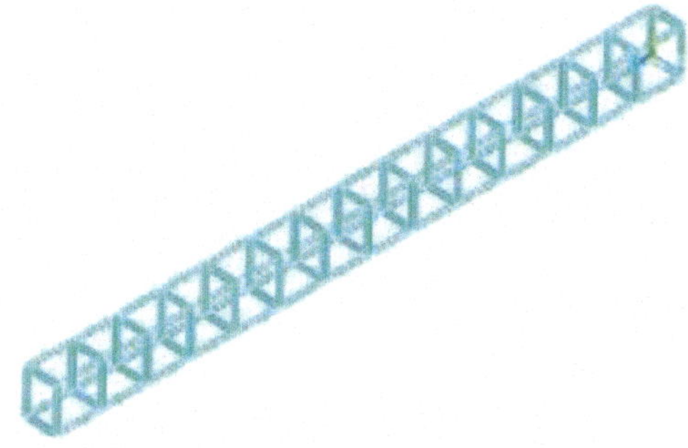

a: Modelling of reinforcement

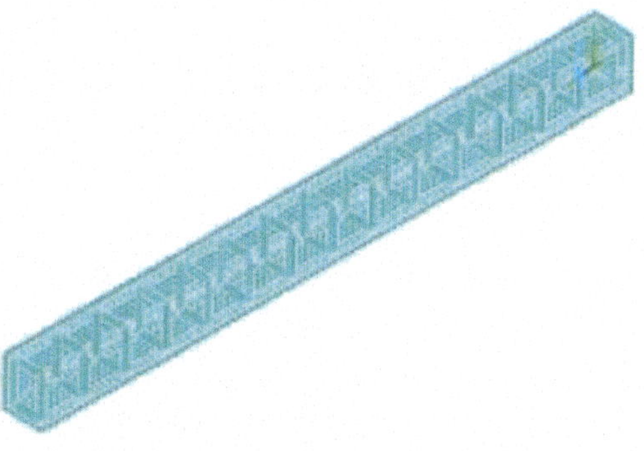

b: Modelling of concrete and reinforcement

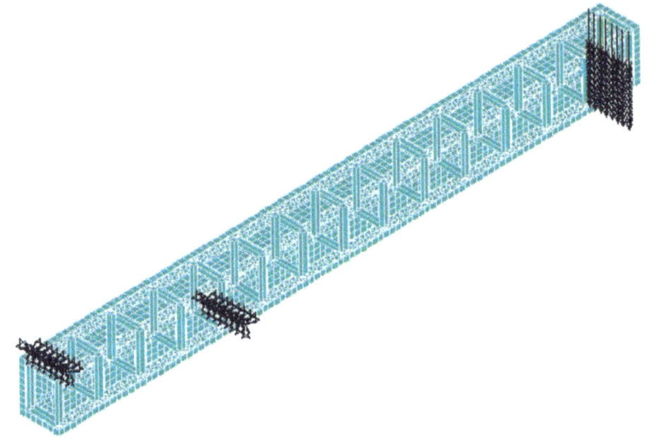

Fig. 4.9 Loads and supports in ANSYS APDL

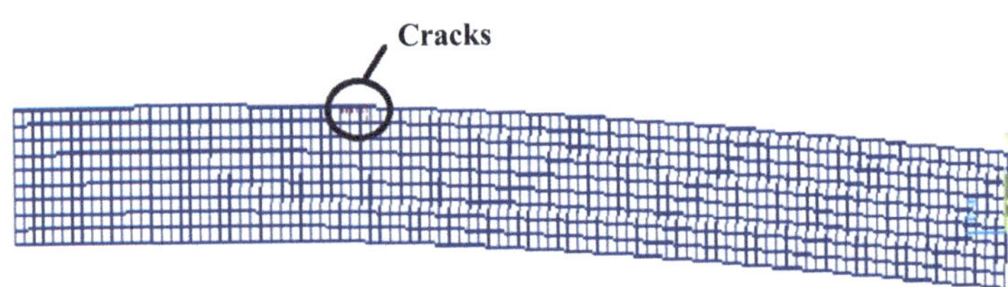

Fig. 4.10 Failure patterns on fencing posts during numerical simulation

The William and Warnke failure model was adopted for concrete. Cracking was assumed to occur when the principal tensile stress exceeded the defined threshold, while crushing was triggered when the principal compressive stress exceeded the yield surface.

- The simulation was terminated when visible cracks appeared on the surface, in accordance with experimental failure definitions.

Validation of Crack Patterns and First Crack Loads
As shown in Fig. 4.10, the crack patterns from the simulation showed good agreement with those observed in experiments, validating the modelling assumptions. The results of the first crack load obtained from numerical simulations are compared with experimental values in Table 4.4.

4.3.2 *Impact Test on Fencing Post*

To simulate the dynamic response of fencing posts under impact loading, numerical investigations were performed. This approach enabled accurate replication of high strain-rate effects, crack initiation, and energy absorption observed in experimental drop-weight tests.

Table 4.4 Comparison of first crack load during the static test

Specimen ID	First crack load (N)		
	Numerical	Experimental	Percentage increase (%)
M20R0	1181	1200	1.61
M20R15	1032	1000	−3.10
M20R0PP0.2	1241	1200	−3.30
M20R15PP0.2	1026	1000	−2.53
M20R0SF0.75	1270	1300	2.36
M20R15SF0.75	1211	1200	−0.91

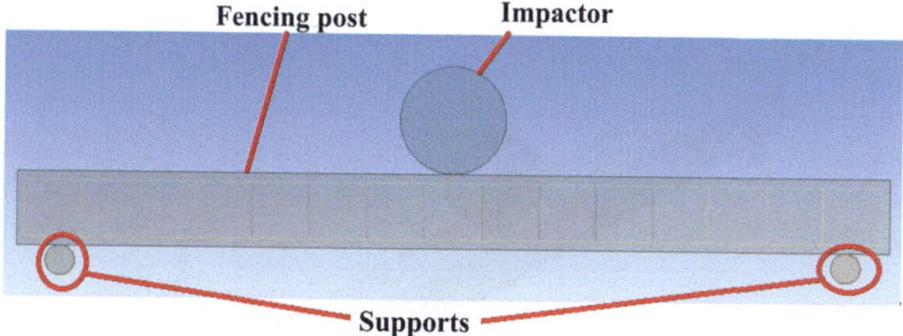

Fig. 4.11 Modelling of reinforcement, fencing post, supports, and impactor

Modelling and Meshing

The geometry of the fencing post, supports, and impactor was recreated in ANSYS Workbench. Reinforcement was modelled as embedded elements within the concrete volume. The impactor was modelled as a rigid body with a mass of 20 kg, matching the experimental setup. Bonded contacts were defined between the post and supports to replicate fixed-end conditions, while a frictionless contact was assumed between the impactor and the top face of the post. Figures 4.11 and 4.12 denote the details of model used for simulation of impact test on fencing post.

Fixed boundary conditions were applied to the base supports. The impactor was assigned an initial velocity, incrementally increased by 50 mm/s per iteration until visible damage or element deletion occurred. Failure was assumed at the point where cracking became visible on the bottom face of the post, in line with experimental definitions.

Validation of Results

The simulation outcomes closely followed the experimental crack patterns, with top face crushing and bottom face tension cracking, as seen in Fig. 4.13. The results of the impact energy absorption capacity for various material variants obtained from numerical analysis are tabulated in Table 4.5, which show strong agreement with the experimental results.

a: Meshing of reinforcement in ANSYS Workbench

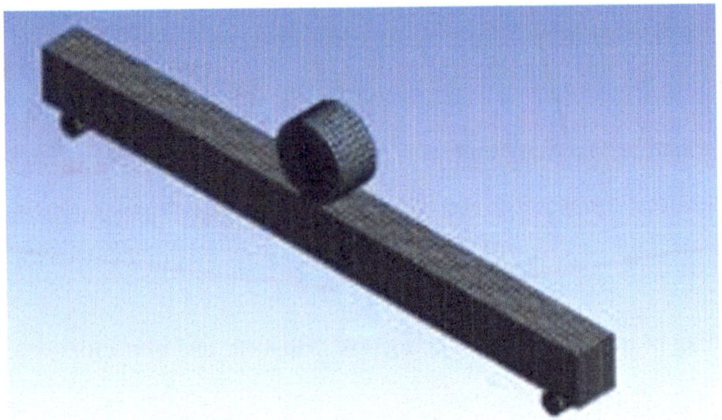
b: Meshing of the entire model

Fig. 4.12 Modelling of fencing post for simulation of impact test. (**a**) Meshing of reinforcement in ANSYS Workbench. (**b**) Meshing of the entire model

Fig. 4.13 Typical failure pattern on fencing posts during numerical simulation of impact test

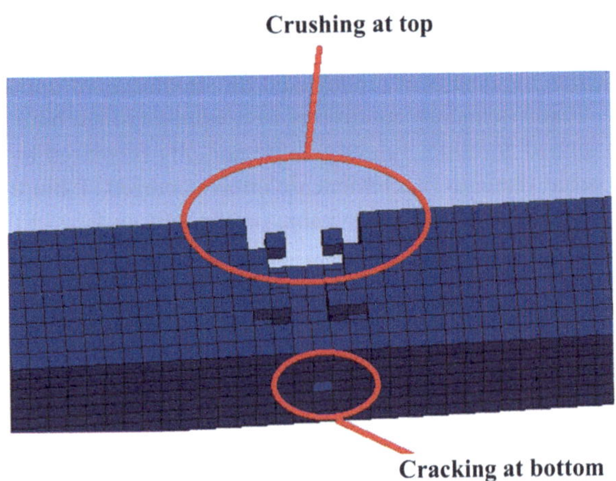

Table 4.5 Numerical studies on impact test

Specimen ID	Velocity at the appearance of cracks (m/s)	Energy absorbed (Nm)
M20R0	2.6	67.6
M20R15	3.0	90.0
M20RPP0.2	3.0	90.0
M20R15PP0.2	3.4	115.6
M20R0SF0.75	3.7	136.9
M20R15SF0.75	4.3	184.9

4.4 Benefit to Cost Ratio

The results from static and impact tests indicated only marginal differences in the static load-carrying capacity of fencing posts made with ordinary concrete, rubcrete, fibre-reinforced concrete, and fibre-reinforced rubcrete. However, a significant variation was observed in their energy absorption capacities under impact loading. To better understand the economic efficiency of each material variant, the energy absorbed per unit cost was evaluated.

Methodology

- The total energy absorbed during the impact test was divided by the corresponding cost of each fencing post specimen to obtain the Energy Absorbed to Cost Ratio (Nm/₹).
- This ratio serves as a practical indicator for selecting materials where impact resistance and cost-efficiency are critical.

Results and Observations

- As shown in Table 4.6, the rubcrete fencing posts offered better energy absorption per rupee spent compared to ordinary concrete posts.
- Polypropylene fibre-reinforced rubcrete fencing posts showed a 63% improvement, while steel fibre-reinforced rubcrete fencing posts demonstrated a 145% improvement over ordinary concrete posts.
- Among all, steel fibre-reinforced rubcrete exhibited the highest energy absorption-to-cost ratio, highlighting its superior performance and economic value.

Sustainability and Resource Conservation

In addition to technical performance and cost-efficiency, the use of steel fibre-reinforced rubcrete fencing posts contributes to resource conservation and sustainability. As per IS:4996-1984, 1 km of fencing requires approximately 334 line posts, each comprising 0.01875 m^3 of concrete.

- Replacing ordinary concrete with steel fibre-reinforced rubcrete leads to:
  - 1.67 kg of fine aggregate saved per post.
  - 0.45 kg of crumb rubber reused per post.

Table 4.6 Energy absorbed to cost ratio

Specimen ID	Cost per sleeper model (₹)	Energy absorbed (Nm)	Energy absorbed to cost ratio (Nm/₹)
R0	323.16	1412.64	4.37
PP0.2	325.36	1569.60	4.82
SF0.75	353.58	2589.84	7.32
R15SF0.75	357.36	3139.20	**8.78**

Therefore, over 1 km of fencing:

- ~550 kg of fine aggregates is conserved.
- ~150 kg of crumb rubber is diverted from landfills and reused.

Such material savings not only reduce dependency on fine aggregates but also address the issue of rubber waste, making this approach a sustainable and environmentally responsible alternative to conventional fencing solutions.

4.5 Summary

This chapter presented a comprehensive investigation into the performance of fencing posts fabricated using various concrete compositions—namely, ordinary concrete, rubcrete, fibre-reinforced concrete, and fibre-reinforced rubcrete. Both experimental and numerical studies were carried out to evaluate their structural response under static and impact loading conditions.

Experimental investigations revealed that all fencing post variants satisfied the minimum static load requirements prescribed by IS: 4996-1984. However, rubcrete and its fibre-reinforced counterparts demonstrated significantly enhanced energy absorption under impact, without compromising structural integrity.

Numerical simulations using ANSYS Mechanical APDL and ANSYS Workbench effectively captured the crack propagation and load-response behaviour observed in physical tests. The close correlation between experimental and numerical results affirmed the validity of the modelling approach and reinforced confidence in the use of simulation for predictive design.

A cost-to-energy absorption analysis further highlighted the superiority of fibre-reinforced rubcrete variants: especially those incorporating steel fibres, not only in mechanical performance but also in economic efficiency. Among all the material types evaluated, steel fibre-reinforced rubcrete emerged as the most optimal solution, offering maximum impact resistance for every rupee spent.

Importantly, the adoption of rubcrete and its fibre-reinforced forms offers tangible environmental benefits. The substitution of fine aggregates with crumb rubber promotes the reuse of waste tyres and reduces the depletion of natural resources. The results of the investigations demonstrated that for every kilometre of fencing using steel fibre-reinforced rubcrete posts, approximately 550 kg of fine aggregates could be conserved and 150 kg of crumb rubber could be reused, underlining the material's potential for sustainable infrastructure applications.

Overall, this chapter affirms that fibre-reinforced rubcrete fencing posts are a structurally viable, cost-effective, and environmentally sustainable alternative to ordinary concrete fencing solutions.

For further details on the application of fibre-reinforced rubcrete on concrete fencing posts, the readers could also refer the journal published by the authors [1].

References

1. A. Raj, P. Nagarajan, S. Aikot Pallikkara, Application of Fiber-reinforced rubcrete in fencing posts. Pract. Period. Struct. Des. Constr. **25**, 04020037 (2020). https://doi.org/10.1061/(asce)sc.1943-5576.0000512
2. Bureau of Indian Standards, Specification for reinforced concrete fencing post. IS: 4996-1984, New Delhi, India (1984)

Chapter 5
Crash Barriers and Kerbs

5.1 Introduction

Ensuring the safety of road users has become a matter of increasing urgency with the rapid expansion of vehicular traffic and highway infrastructure. Among the critical components serving this function are concrete crash barriers and kerbs, both of which are subjected to sudden and often severe impact forces. The ability of these structural elements to absorb energy efficiently and retain their integrity under impact is a decisive factor in preventing fatalities and minimising damage during accidents.

Crash barriers, typically positioned along medians and roadside edges, are designed to contain or redirect errant vehicles. Their effectiveness hinges not only on geometric configuration but also on the inherent material properties of the concrete used. With escalating speeds, axle loads, and transport tonnages, ordinary concrete barriers are increasingly being pushed to their performance limits. Figure 5.1 presents a scenario of failure of concrete crash barriers due to collision from a vehicle.

In this context, the adoption of steel fibre-reinforced rubcrete and polypropylene fibre-reinforced rubcrete presents a promising avenue for enhancing the impact resistance and energy absorption capabilities of crash barriers. However, prior to their implementation in real-world scenarios, rigorous laboratory-scale testing and numerical validation are needed to ascertain their suitability.

On the other hand, kerbs, though traditionally perceived as marginal structural components, often experience vehicular contact, particularly in urban settings (Fig. 5.2). The impact loads imposed by tyres, especially from heavy vehicles can lead to localised damage, cracking, or even failure of kerbs over time. While experimental testing on kerbs can be logistically challenging, numerical simulations offer a reliable alternative to study their performance under impact, enabling parametric evaluation across multiple material variants.

A. Raj et al., *Rubcrete and Its Applications in Structures Subjected to Impact Loading*, SpringerBriefs in Applied Sciences and Technology,
https://doi.org/10.1007/978-981-95-6315-9_5

Fig. 5.1 Failure of concrete crash barriers due to impact loads

Fig. 5.2 Damaged kerb

This chapter presents a comprehensive study on the performance of crash barriers and kerbs under impact loading conditions. The first part encompasses both experimental investigations and numerical modelling of concrete crash barriers, including prototype simulations. The second part exclusively addresses numerical studies on kerbs, evaluating six different material variants for their energy absorption capacities and resistance to failure. Across both cases, the focus is on assessing the suitability of advanced concrete composites, particularly those incorporating crumb rubber and fibres as viable materials for modern road safety infrastructure.

5.2 Experimental Investigations on Crash Barriers

Crash barriers are indispensable safety features on modern highways, functioning as containment structures to mitigate the consequences of vehicular deviations from the travel path. With increasing vehicular tonnage and speeds, the demand for higher impact resistance and energy absorption in crash barriers has become paramount. Therefore, the suitability of steel fibre-reinforced rubcrete and polypropylene fibre--reinforced rubcrete for use in crash barriers has been experimentally assessed through scaled impact testing.

A one-third scale model of the New Jersey crash barrier was employed to facilitate controlled laboratory experimentation [1]. The barrier geometry, adapted from standard profiles, is illustrated in Fig. 5.3.

The specimens were cast using concrete of strength 40 N/mm^2, consistent with the optimised mix proportions detailed in Chap. 2. Six different material variants were considered (Table 5.1):

The reinforcement cage, formwork, and cast crash barrier specimens are shown in Figs. 5.4, 5.5, and 5.6.

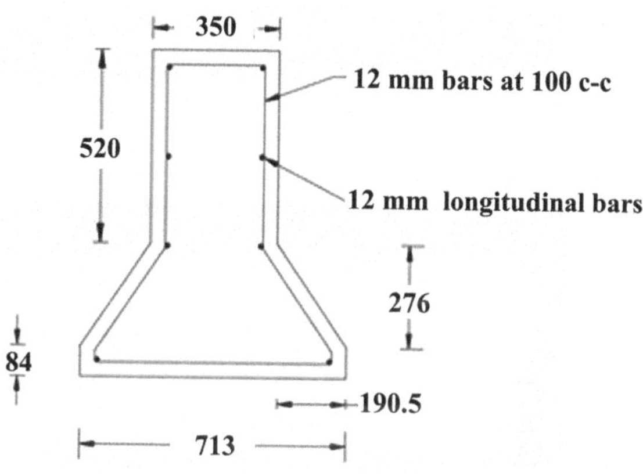

Fig. 5.3 Geometry of the New Jersey crash barrier (all dimensions in mm)

Table 5.1 Number of crash barrier specimens

Specimen I D	Specimens cast
R0	2
R15	2
R0PP0.2	2
R15PP0.2	2
R0SF0.75	2
R15SF0.75	2

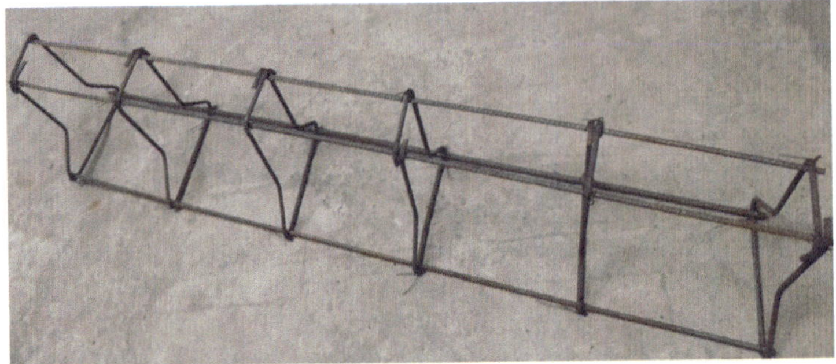

Fig. 5.4 Reinforcement cage for the crash barrier

Fig. 5.5 Mould for casting the crash barrier

Fig. 5.6 Cast crash barrier specimen

Impact Testing Setup

Given the prohibitive cost and logistical complexity of full-scale vehicle crash testing, a pendulum-type impact testing apparatus was fabricated. The device allowed for controlled impact simulations using a swinging mass of 50 kg attached to a 1.5 m long pendulum rod, with an effective drop height of 500 mm.

The test configuration, designed to impart repeated impacts until visible cracking occurred, is presented in Fig. 5.7.

Crash barrier specimens were restrained using bolted steel frames during testing. The energy absorbed upon failure was calculated based on the number of effective blows delivered prior to visible cracking (Table 5.2).

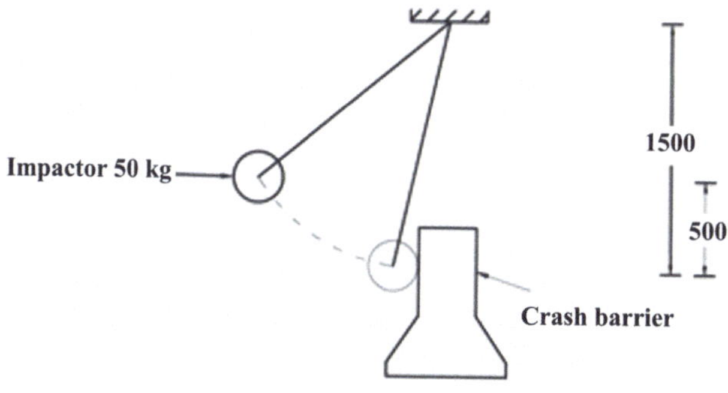

a: Schematic diagram for impact testing of crash barriers

b: Impact testing of crash barriers

Fig. 5.7 Pendulum-type impact testing of crash barrier. (**a**) Schematic diagram for impact testing of crash barriers. (**b**) Impact testing of crash barriers

Table 5.2 Energy absorbed during experimental testing of crash barrier models

Specimen ID	No. of blows	Energy absorbed (kNm)	Percentage increase
R0	4	0.981	–
R15	6	1.471	49.95
R0PP 0.2	5	1.226	24.97
R15 PP0.2	7	1.716	74.92
R0SF0.75	8	1.962	100
R15SF0.75	13	3.188	224.97

Fig. 5.8 Cracks in crash barriers subjected to impact test

The failure patterns observed post-testing are shown in Fig. 5.8.

From the above experiments the following can be observed.

- The incorporation of crumb rubber significantly enhanced the energy absorption capacity due to its inherent deformability and damping characteristics.
- Polypropylene fibres, through their bridging action, delayed crack initiation and propagation, thereby increasing the energy required to compromise the concrete matrix.
- In the case of steel fibre-reinforced rubcrete, the interaction of rigid fibre and flexible crumb rubber particles resulted in the highest energy absorption capacity, confirming their superior suitability for high-impact applications such as crash barriers.
- The use of fibres, especially steel fibres contributed to post-crack load transfer mechanisms, allowing for more sustained energy dissipation before failure.

5.3 Numerical Investigations on Crash Barriers

To complement experimental findings and extend the evaluation to prototype structural elements, numerical simulations were carried out using the Explicit Dynamics module of ANSYS Workbench. These simulations allowed for the nonlinear dynamic analysis of crash barriers under high-velocity impact loads, enabling the assessment of both scaled models and full-scale prototypes constructed with fibre--reinforced rubcrete.

5.3.1 Modelling Approach

The geometry of structure, reinforcement, impactor, and supports was generated using the ANSYS Design Modeler. The reinforcement was embedded within the concrete using the "reinforcement" interaction type, which defines the line elements (steel bars) as fully bonded within the concrete solid body. Material properties adopted for concrete and fibres are given in Chap. 2.

Meshing was carried out after performing mesh sensitivity analysis, using a combination of quadrilateral and triangular elements. The final mesh for the scale model comprised 17,404 elements and 23,000 nodes. The details of modelling reinforcement of crash barriers can be seen from Fig. 5.9. The finite element model of the crash barrier is shown in Fig. 5.10.

Boundary conditions were applied to restrict all translations and rotations at the supports. The contact between the impactor and the concrete surface was considered frictionless, and failure criteria for concrete was defined using the principal stress failure criterion. Analysis was carried out by incrementally increasing the impactor velocity (in steps of 50 mm/s) until failure was observed at the far side of the barrier (Fig. 5.11).

The energy absorbed was calculated from the kinetic energy of the impactor at the point of failure (Table 5.3).

5.3.2 Validation of Numerical Model

A comparison of the experimental and numerical results (Table 5.4) for the scaled crash barrier models revealed excellent correlation, validating the accuracy of the simulation model.

This close agreement affirms the reliability of the numerical model, allowing for its use in extended simulations, such as altered support conditions and prototype--scale modelling.

5.3.3 Simulation of Fixed Support and Prototype Crash Barriers

With the validated numerical model for crash barriers in place, further simulations were conducted to explore the effects of support conditions and to assess the behaviour of prototype-scale crash barriers made using fibre-reinforced rubcrete and its variants.

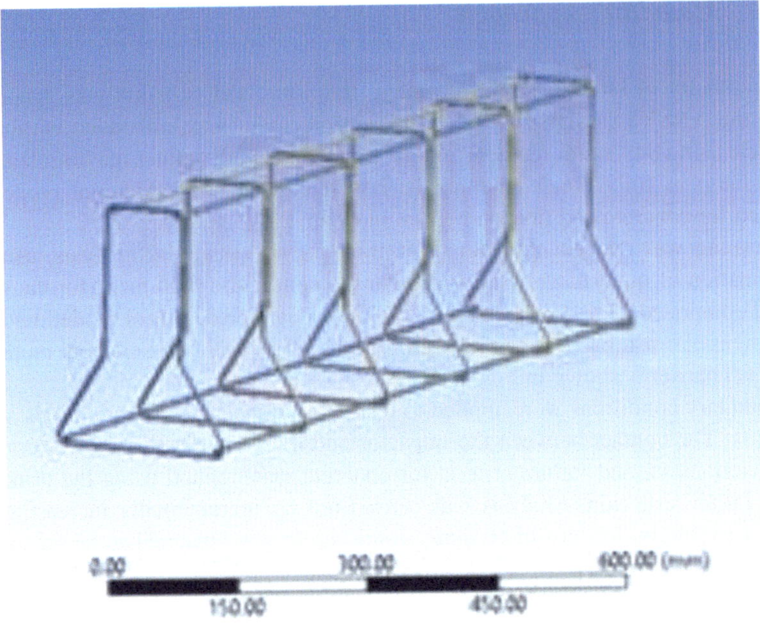

5.9 a: Reinforcement for the scale model of the crash barrier

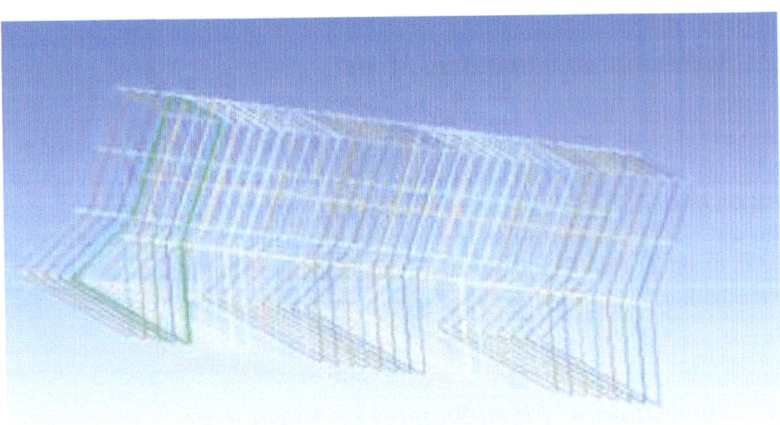

5.9 b: Reinforcement for the prototype of the crash barrier

Fig. 5.9 Modelling of reinforcement. (**a**) Reinforcement for the scale model of the crash barrier. (**b**) Reinforcement for the prototype of the crash barrier

Effect of Support Conditions: Fixed-Base Simulation

In real-world applications, crash barriers are typically anchored to the base using rigid fixings. To replicate this, a fixed-base simulation was performed using the previously validated model. The concrete base was assigned fixed displacement boundary conditions, and frictionless contact was assumed between the impactor and concrete surface.

Fig. 5.10 Crash barrier models after meshing. (**a**) Crash barrier scale model after meshing. (**b**) Prototype crash barrier after meshing

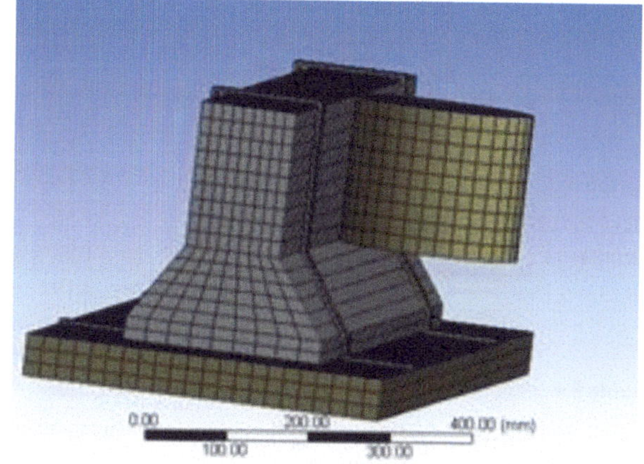

a: Crash barrier scale model after meshing

b: Prototype crash barrier after meshing

Cracking at far side

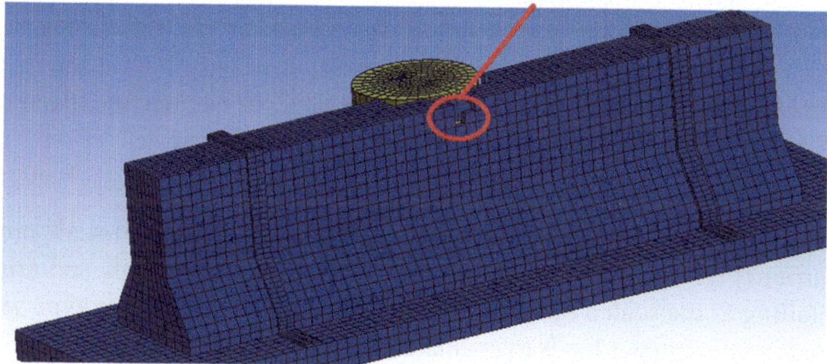

Fig. 5.11 Failure modes of crash barrier models

Table 5.3 Energy absorbed by crash barriers

Specimen ID	Velocity at which cracks appeared in m/s	Energy absorbed (kNm)	Percentage increase
R0	6.20	0.961	–
R15	7.40	1.369	42.46
R0PP 0.2	6.80	1.156	20.29
R15 PP0.2	7.90	1.56	62.36
R0SF0.75	8.90	1.98	106.06
R15SF0.75	11.30	3.192	232.18

Table 5.4 Comparison of energy absorption capacities

Specimen I D	Energy absorption of scale-down model		
	Experimental (Nm)	Numerical (Nm)	Percentage increase
R0	981	961	2
R15	1471	1369	7
R0PP0.2	1226	1156	6
R15PP0.2	1716	1560	9
R0SF0.75	1962	1980	−1
R15SF0.75	3188	3192	−0.1

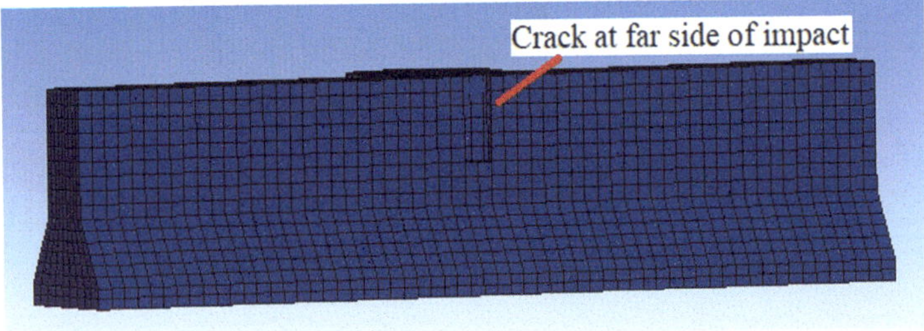

Fig. 5.12 Typical failure pattern of fixed-base crash barrier model (for scaled crash barrier)

Meshing was carried out using the same mesh convergence parameters, yielding 14,760 nodes and 12,030 elements.

The simulation revealed that the fixed-base support condition enhanced the energy absorption capacity of the barrier marginally (roughly 7%), when compared to the simply supported counterpart (Fig. 5.12). Hence it can be concluded that the support condition does not affect the capacity of the crash barrier.

Full-Scale Prototype Simulation

A detailed three-dimensional simulation was carried out on the full-scale prototype crash barrier (Fig. 5.10b), adopting the same cross-sectional profile and reinforcement detailing as the scaled model but scaled appropriately. The prototype was subjected to dynamic impact loads to evaluate its real-world performance.

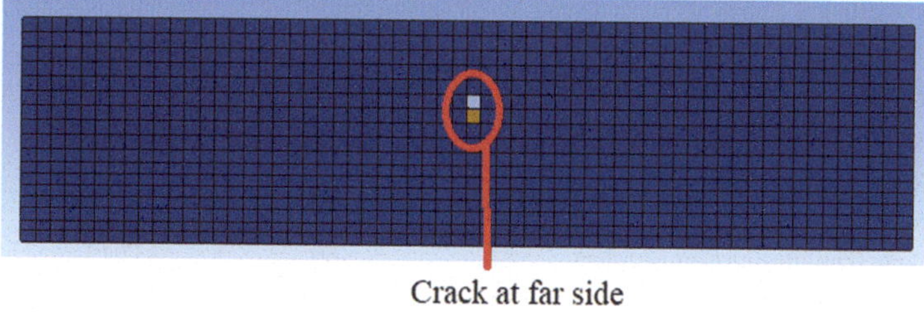

Fig. 5.13 Crack formation in prototype crash barrier (ANSYS simulation output)

Table 5.5 Energy absorption during impact on prototype crash barriers

Sl.No	Specimen ID	Mass (kg)	Velocity (m/s)	Energy absorbed (kNm)	Percentage increase
1	R0	1500.0	29.600	657.12	–
2	15	1500.0	36.000	972.00	47.92
3	R0PP0.2	1500.0	34.000	867.00	31.94
4	R15PP0.2	1500.0	39.450	1167.23	77.63
5	R0SF0.75	1500.0	40.750	1245.42	89.53
6	R15SF0.75	1500.0	48.000	1728.00	162.97

- The impactor mass was taken as 1500 kg, simulating vehicular impact scenarios.
- The boundary condition was fixed at the base to mimic actual field installation.
- Mesh convergence studies led to the generation of 15,456 nodes and 11,423 elements.
- A mix of quadrilateral and triangular elements was used for accuracy and computational stability.

The energy absorbed by the specimens at the time of failure (Fig. 5.13) are given in Table 5.5.

The results clearly show that steel fibre-reinforced rubcrete (R15SF0.75) exhibited the highest energy absorption, validating its potential for application in highway safety infrastructure.

5.3.4 Compliance with Codal Requirements and Suitability of Variants

To ensure practical applicability, the energy absorption capacities of prototype crash barriers developed using various concrete formulations were assessed against the minimum requirements specified by governing standards. This evaluation provides insights into the feasibility of employing sustainable concrete variants, particularly fibre-reinforced rubcrete, in highway safety systems.

Table 5.6 Collision energy of typical vehicles on Indian roads

Vehicle type	Tonnage (tonnes)	Max speed (km/h)	Energy during collision (kNm)
Single axle (1 tyre)	3	60	48.17
Single axle (2 tyres)	7.5	60	120.42
Single axle (4 tyres)	11.5	60	184.64
Tandem axle (rigid/trailer/semi-trailer)	21	60	337.17
Tandem axle (puller tractors with hydraulic/pneumatic trailers)	28.5	60	457.58
Tri-axle vehicles	27	60	433.5
Modular hydraulic trailers (axle row with 4 tyres each)	18	60	289
Rigid vehicles (gross weight)	49	60	786.72
Semi-articulated truck-trailers (gross weight)	55	60	883.06

Comparison with Vehicle Collision Energy Standards

Crash barriers are required to withstand impacts from vehicles of varying configurations and weights, travelling at designated highway speeds. The energy levels during such collisions were computed based on data from the Press Information Bureau (2018) [2], which specifies gross vehicle weights and typical velocity limits (Table 5.6).

The energy absorption capacities of the prototype crash barriers obtained from numerical simulations (Sect. 5.3.3) were compared with these benchmark values to assess their adequacy.

- The ordinary concrete crash barrier (R0) withstood 657.12 kNm, which falls short of containing vehicles weighing over 49 tonnes.
- In contrast, the steel fibre-reinforced rubcrete barrier (R15SF0.75) exhibited a remarkable energy absorption capacity of 1728.00 kNm, which is approximately 2.6 times the collision energy of a 55-tonne articulated trailer moving at 60 km/h.

This indicates that R15SF0.75 can effectively absorb the impact of even the heaviest vehicles allowed on Indian highways, offering a robust and reliable alternative to conventional barriers.

Suitability of Material Variants

The assessment of all six material variants considered for the prototype barriers leads to the following observations:

- Rubcrete (R15) and polypropylene fibre-reinforced rubcrete (R15PP0.2) offer moderate improvements in impact performance and are suitable for medium-duty highways.
- Steel fibre-reinforced concrete (R0SF0.75) and steel fibre-reinforced rubcrete (R15SF0.75) outperform all other variants by a significant margin in energy dissipation capacity.

- R15SF0.75, combining the toughness of crimped steel fibres and the damping characteristics of crumb rubber, emerges as the most effective material variant.

Hence, based on both experimental and numerical evidence, steel fibre-reinforced rubcrete is identified as the preferred choice for crash barrier construction in high-speed and heavy-traffic zones.

5.4 Benefit-to-Cost Ratio

Energy Absorption-to-Cost Analysis

In assessing the viability of alternative materials for crash barriers, it is essential to evaluate not only their mechanical performance but also their cost-effectiveness. The energy absorption-to-cost ratio offers a pragmatic metric, reflecting the amount of impact energy a crash barrier can absorb per unit cost of construction (Table 5.7).

From this analysis, it is evident that:

- Rubcrete crash barriers (R15) provide a 46% improvement in energy-to-cost ratio over ordinary concrete (R0).
- Polypropylene fibre-reinforced rubcrete (R15PP0.2) improves this ratio by 68%.
- Steel fibre-reinforced rubcrete (R15SF0.75) offers a staggering 183% increase, making it the most efficient variant from both structural and economic perspectives.

The adoption of steel fibre-reinforced rubcrete is not merely a technical enhancement: it is a strategic response to the challenges of material scarcity and environmental degradation.

- For every kilometre of crash barrier constructed using R15SF0.75, approximately:
  - 48 tonnes of natural fine aggregates can be conserved.
  - 14 tonnes of discarded tyre rubber can be productively repurposed.

This dual benefit underscores the ecological value of steel fibre-reinforced rubcrete:

- It mitigates the over-extraction of river sand, preserving delicate ecosystems.
- It offers a high-value application for end-of-life tyres, reducing the burden on landfills and the associated environmental hazards.

Table 5.7 Energy absorption-to-cost ratio for various crash barrier variants

Specimen ID	Energy absorbed (Nm)	Material cost per barrier (₹)	Energy absorption-to-cost ratio (Nm/₹)
R0	981	2260	0.434
R15	1471	2319.6	0.634
R15PP0.2	1716	2347	0.731
R15SF0.75	3188	2594	1.23

Thus, the use of fibre-reinforced rubcrete, particularly R15SF0.75, aligns with the principles of sustainable infrastructure development, enabling technical excellence while advancing resource efficiency and environmental stewardship.

Summary of Crash Barrier Performance Evaluation

The experimental and numerical investigations conducted on one-third scale models and prototype crash barriers have led to several key insights regarding the impact resistance and viability of alternative concrete materials.

- **Impact Energy Absorption**:
  - Steel fibre-reinforced concrete and rubcrete barriers exhibited significant improvements in energy absorption capacity.
  - Compared to ordinary concrete (R0), the steel fibre-reinforced rubcrete barrier (R15SF0.75) demonstrated:

    224% higher energy absorption in laboratory-scale tests.
    163% greater energy absorption in prototype numerical simulations.

- **Numerical Validation**:
  - The finite element models developed in ANSYS Workbench accurately captured the behaviour observed in physical testing.
  - A comparison between experimental and numerical results showed less than 10% deviation, confirming the model's reliability.

- **Performance Benchmarking**:
  - Prototype simulations revealed that ordinary concrete barriers may not withstand vehicular impacts from heavy-duty trucks and articulated trailers as per recent transportation norms.
  - In contrast, R15SF0.75 barriers exceeded the impact energy threshold required to contain vehicles up to 55 tonnes moving at 60 km/h.

- **Economic and Environmental Viability**:
  - Steel fibre-reinforced rubcrete offered the highest energy absorption to cost ratio among all tested variants.
  - Its implementation can lead to substantial material savings, including:

    48 tonnes of fine aggregates saved.
    14 tonnes of scrap tyre rubber reused per kilometre of crash barrier constructed.

In conclusion, R15SF0.75 (steel fibre-reinforced rubcrete) has emerged as the most promising material for impact-resistant crash barriers, striking an optimal balance between mechanical performance, cost-effectiveness, and sustainability.

5.5 Kerbs

Concrete kerbs play a crucial role in road infrastructure by acting as physical boundaries that control vehicular movement and define pavement edges. While they primarily serve to guide traffic and prevent encroachment, they are also subjected to low-velocity impact loads, particularly from vehicles grazing or mounting them during turns or emergency manoeuvres. Although not designed for high-energy collisions like crash barriers, the structural performance of kerbs under repeated low-intensity impacts is essential for maintaining road safety and serviceability.

5.6 Numerical Investigations on Kerbs

Given the practical challenges and costs of conducting full-scale experimental studies on kerbs under impact, a numerical approach was adopted to assess their performance.

The kerbs were modelled and analysed in ANSYS Workbench using the Explicit Dynamics module, which is well-suited for simulating short-duration, high strain-rate loading conditions such as vehicle impact (Figs. 5.14, 5.15, and 5.16).

Model Geometry and Simulation Details
To accurately simulate the response of concrete kerbs under vehicular impact, a detailed finite element model was developed in the Design Modeller module of ANSYS Workbench. The kerb model incorporated both the concrete geometry and embedded reinforcement, along with an impactor representing a moving vehicle tyre.

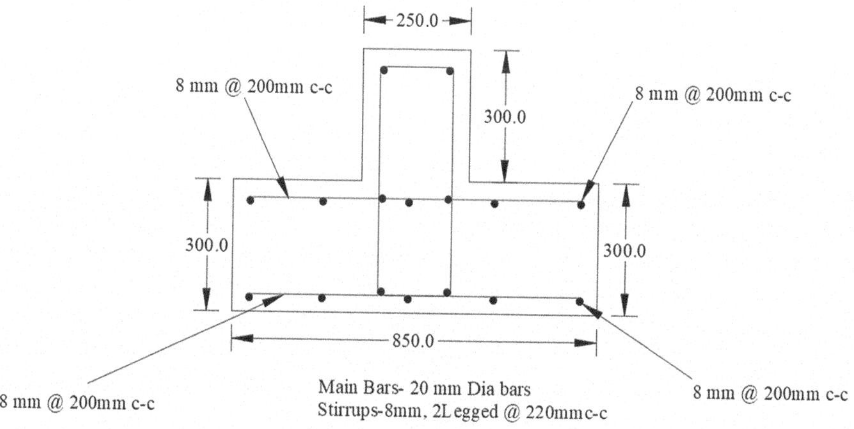

Fig. 5.14 Details of the kerb (all dimensions in mm)

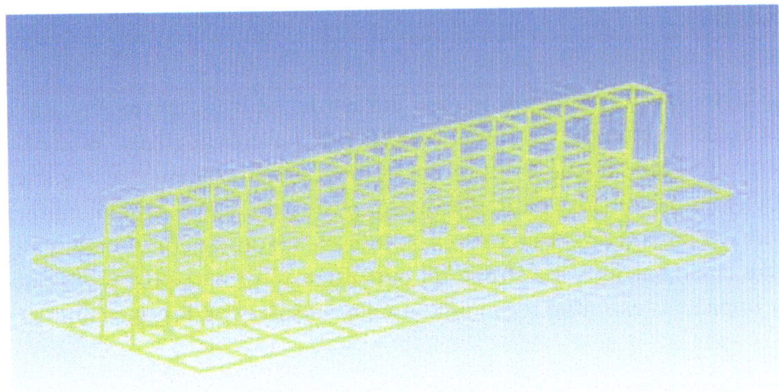

Fig. 5.15 Reinforcement in kerb

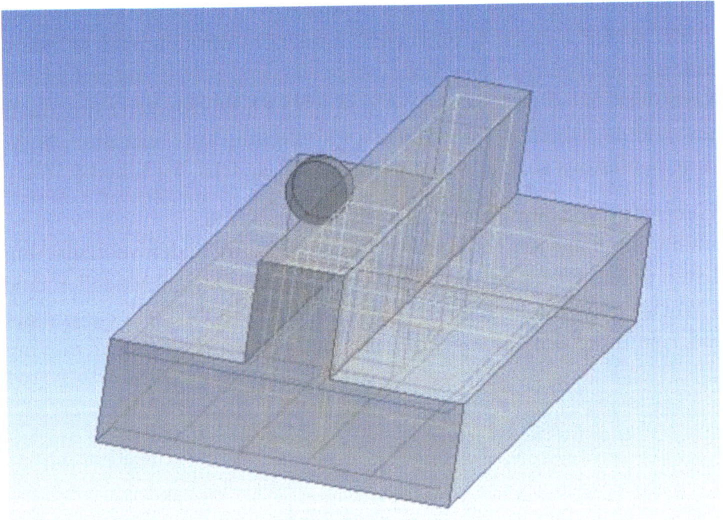

Fig. 5.16 3D numerical model of kerb with impactor

Kerb Geometry and Material

The kerb was modelled using concrete of strength 20 N/mm^2 with dimensions and reinforcement details derived from typical road infrastructure specifications. The concrete kerb section was reinforced with longitudinal bars and stirrups, consistent across all six material variants. Crumb rubber and fibres (steel or polypropylene) were incorporated by modifying the material properties to reflect the corresponding mix characteristics.

Impactor Configuration

The impactor was modelled as a rigid body with a mass of 20.35 tonnes, selected to replicate the wheel-load equivalent of a passenger bus, based on standard traffic impact studies. The contact between the impactor and the kerb was assumed to be frictionless to simulate worst-case impact transfer conditions.

Boundary Conditions

The base slab supporting the kerb was assigned fixed displacement boundary conditions, effectively simulating its attachment to a road base. Reinforcement was embedded using the "reinforcement" interaction type available in ANSYS, ensuring proper load transfer between steel and concrete. Crack propagation was captured using element deletion based on the maximum principal stress failure criterion, which effectively removed elements once their failure threshold was reached.

Meshing

A refined mesh was developed after conducting mesh convergence analysis to ensure the accuracy of results. The final model consisted of 173,259 nodes and 154,436 elements, composed of tetrahedral and hexahedral elements for optimal accuracy and computational efficiency.

Loading

Impact loading was applied by incrementally increasing the velocity of the impactor until visible failure (crack formation) was detected at the far face of the kerb. The onset of failure was recorded for each material variant, and the corresponding energy absorbed was computed. This procedure ensured consistency across all simulations and provided a comparative measure of the energy absorption capacity of each concrete mix.

Results and Discussion

The numerical simulations yielded valuable insights into the energy absorption characteristics of kerbs made using various concrete formulations. For each material variant, the velocity at which cracking was initiated and the corresponding energy absorbed were noted (Fig. 5.17). These results are summarised in Table 5.8.

The following observations can be made from the above results:

- Rubcrete kerbs (R15) demonstrated a 44% increase in energy absorption compared to ordinary concrete (R0), confirming the beneficial role of crumb rubber in enhancing impact resistance.

Fig. 5.17 Cracks in kerbs during numerical simulation

Table 5.8 Energy absorbed by kerbs under impact

Specimen I D	Velocity at cracking (m/s)	Energy absorbed (10^6 kNm)	Percentage increase
R0	415	1.75	0
R15	498	2.52	44
R0PP0.2	483	2.37	35.46
R15PP0.2	504	2.59	47.49
R0SF0.75	600	3.66	109.03
R15SF0.75	685	4.78	172.45

Table 5.9 Number of impacts till failure

Specimen ID	Number of impacts to failure
R0	90,435
R15	130,227
R0PP	122,500
R15PP	133,384
R0SF	189,036
R15SF	246,390

- The presence of polypropylene fibres resulted in an appreciable increase in energy absorption, with R0PP0.2 and R15PP0.2 kerbs showing 35.46% and 47.49% improvements, respectively.
- Steel fibre incorporation led to the most significant enhancement. R0SF0.75 and R15SF0.75 kerbs exhibited 109% and 172.45% higher energy absorption, respectively, over plain concrete kerbs. This demonstrates the superior crack-bridging capacity of steel fibres during high-impact events.

The number of impacts each variant could withstand, based on a standard energy threshold (energy per impact for a 20.35-tonne vehicle at 5 km/h), is shown in Table 5.9.

These results indicate:

- A clear enhancement in the fatigue life of kerbs when reinforced with either fibres or crumb rubber.
- R15SF0.75 kerbs withstood over 2.7 times more impacts than conventional kerbs, indicating excellent long-term performance under repeated vehicle impacts.
- Even polypropylene fibre-reinforced rubcrete variants show more than 47% improvement in impact resistance over plain concrete.

These results underscore the effectiveness of fibre reinforcement systems, specifically steel fibre-reinforced rubcrete, in elevating both the resilience and durability of concrete kerbs subjected to impact loading.

5.7 Summary

This chapter presented a detailed evaluation of the impact resistance of concrete crash barriers and kerbs through experimental and numerical investigations. The results highlighted the considerable advantages of incorporating crumb rubber and fibres into concrete, particularly in the context of structures subjected to vehicular impact.

Crash Barriers

- Experimental studies on one-third scaled New Jersey crash barrier models showed that replacing part of the mineral aggregates with crumb rubber improved energy absorption capacity significantly.
- The inclusion of steel fibres (0.75%) and polypropylene fibres (0.2%) further enhanced the impact resistance.
- Among all variants, steel fibre-reinforced rubcrete performed the best, absorbing 224% more energy than ordinary concrete barriers during scaled experiments.
- Numerical simulations using ANSYS Workbench validated these findings. The model accurately replicated the experimental observations with less than 10% variation in energy absorption.
- Prototype simulations confirmed the scalability and field applicability of the findings. The steel fibre-reinforced rubcrete prototype absorbed up to 163% more energy than conventional barriers, exceeding requirements for resisting impacts from vehicles up to 55 tonnes at 60 km/h.
- Sustainability impact: For every kilometre of crash barrier installed using this variant, approximately 48 tonnes of natural sand and 14 tonnes of waste tyre rubber could be saved and reused, respectively.

Concrete Kerbs

- Only numerical investigations were carried out on concrete kerbs using the explicit dynamics module in ANSYS Workbench.
- Rubcrete and fibre-reinforced rubcrete variants provided substantial gains in impact resistance:
  - Rubcrete increased energy absorption by 44%.
  - Polypropylene fibre-reinforced rubcrete showed a 47.5% improvement.
  - Steel fibre-reinforced rubcrete offered the highest enhancement at 172.45%, withstanding 246,390 impacts, compared to 90,435 for ordinary concrete.
- These results suggest that the hybrid combination of crumb rubber and steel fibres provides an optimal balance of ductility, toughness, and sustainability.

Inferences

- The adoption of fibre-reinforced rubcrete, particularly with steel fibres, offers a technologically viable, economically justifiable, and environmentally sustainable solution for infrastructure exposed to impact loads.

- These materials can significantly prolong service life, reduce maintenance, and promote circular economy through the reuse of waste rubber.
- The findings strongly support the use of these materials in field-scale applications, especially in high-speed, high-load traffic environments.

The readers who want to explore in detail can refer the journal published by the authors [3].

5.8 Final Remarks

This book presents a detailed investigation into a novel construction material, fibre-reinforced rubcrete, with the objective of identifying its suitability for critical structural applications. Four infrastructure components that are frequently subjected to impact loading were focussed upon: railway sleepers, fencing post, crash barriers, and kerbs.

Comprehensive testing was conducted on normal, medium, and high-strength concrete to evaluate performance variations. Among the mixes studied, an optimum formulation was identified: 15% of the volume of fine aggregates to be replaced with crumb rubber, combined with 0.75% steel fibres occupying the total volume of concrete. This combination provided the most favourable balance between economic viability and enhanced impact resistance, without compromising essential structural performance requirements.

The adoption of steel fibre-reinforced rubcrete in the above applications offers considerable material savings while promoting resource conservation. When used in place of ordinary concrete sleepers, approximately 14 tonnes of fine aggregates can be saved and 4 tonnes of crumb rubber can be reused for every kilometre of railway track laid. For crash barriers, this substitution results in savings of about 48 tonnes of fine aggregates and the utilisation of 14 tonnes of worn-out tyre rubber per kilometre. Similarly, the application in fencing posts enables the conservation of nearly 0.5 tonnes of natural fine aggregates and the reuse of 0.15 tonnes of crumb rubber for every kilometre of fencing installed. Such quantifiable environmental gains demonstrate the tangible benefits of integrating waste-derived materials into mainstream infrastructure.

Most importantly, this research has shown the potential of converting discarded tyre rubber into a structurally valuable material. This approach not only addresses the environmental burden of waste tyre disposal but also enhances the resilience and service life of key infrastructure components.

Looking ahead, the promising results obtained here open the door to exploring additional applications of fibre-reinforced rubcrete in both structural and non-structural domains. Future research should extend into full-scale field validations and investigate its adaptability to other impact-prone structures, thereby paving the way for broader acceptance and adoption of this sustainable, high-performance construction material.

References

1. A. Raj, P. Nagarajan, S. Aikot Pallikkara, Application of Fiber-reinforced rubcrete for crash barriers. J. Mater. Civ. Eng. **32**, 04020358 (2020). https://doi.org/10.1061/(ASCE)MT.1943-5533.0003454
2. Press Information Bureau of, India G of, India M of RT& H, The Ministry of Road Transport & Highways has issued a notification increasing permissible truck axle load. PIB Delhi New Delhi, India (2018)
3. A. Raj, P. Nagarajan, S.A. Pallikkara, Performance of fiber-reinforced rubcrete curbs subjected to impact loads. Pract. Period. Struct. Des. Constr. **27**(3) (2022). https://doi.org/10.1061/(ASCE)SC.1943-5576.0000707

Appendix

This section presents the procedure to calculate the proportions of materials required to make 1 m^3 of rubcrete and fibre-reinforced rubcrete for a concrete strength of 60 N/mm^2. The same procedure can be used to get concrete of different strength.

Calculation of material proportions for 1 m^3 concrete of strength 60 N/mm^2

Case 1: Ordinary Concrete

Stipulations for the Mix

Cement	: Ordinary Portland Cement (OPC)
Supplementary cementitious material	: Metakaolin
Coarse Aggregates	: Crushed stones
Nominal size of coarse aggregates	= 6 mm
Fine Aggregates	: Manufactured sand

Specific Gravity of Materials

Cement (G_C)	= 3.14
Metakaolin (G_M)	= 2.6
Coarse aggregates (G_{CA})	= 2.79
Fine aggregates (G_{FA})	= 2.675
Crumb rubber (G_{CR})	= 0.65
High range water reducer (G_{HR})	= 1.08
Viscosity modifying agent (G_{VMA})	= 1.02
Steel fibres	= 7.85
Polypropylene fibres	= 0.91

A. Raj et al., *Rubcrete and Its Applications in Structures Subjected to Impact Loading*, SpringerBriefs in Applied Sciences and Technology,
https://doi.org/10.1007/978-981-95-6315-9

Steps Followed for the Determination of Material Quantities

1. Fixing the target strength for the mix (Table 2, IS 10262: 2019 [1])
   Assumed standard deviation (X) for concrete of strength 60 N/mm^2: 5

$$f'_{ck} = f_{ck} + 1.65X$$

$$f'_{ck} = 60 + 1.65 \times 5$$

$$f'_{ck} = 68.25\,\text{N}/\text{mm}^2$$

   Where
   f'_{ck} is the target compressive strength at 28 days.
   f_{ck} is the characteristic compressive strength at 28 days.
2. Selection of water cement ratio, water content, and total cement content
   The water cement ratio for the initial trial of the mix was selected from Table 8 of IS 10262: 2019. The subsequent trials satisfying the required strength and workability criterion led to the finalisation of 0.33, 162 kg and 490 kg as the water cement ratio, water content, and the total cementitious materials for the mix.
3. Fixing the percentage volume of fine aggregates and coarse aggregates
   For the initial trial, the percentage of volume of fine aggregates and coarse aggregates can be determined from Table 10, IS 10262: 2019. The percentage volume of fine aggregates and coarse aggregates obtained was 37% and 63%, respectively.
4. Determination of volume of materials in the concrete mix
   From previous studies, it was understood that if metakaolin is used to replace cement to improve the overall strength and workability of the mix. Metakaolin also helps in reducing shrinkage cracks. The optimum percentage of replacement of metakaolin was 10% by weight of cement [2]. In this context, determination of mass of metakaolin for the exact volume of cement replaced becomes critical.

Mass of cement (M_C)	= 441 kg
Volume of cement (V_C)	$= \frac{M_C}{G_C \times 1000} = \frac{441}{3.14 \times 1000} = 0.140\,\text{m}^3$
Volume of cement replaced by metakaolin (V_M)	$= \frac{490 - 441}{3.14 \times 1000} = 0.0156\,\text{m}^3$
Volume of water (V_W)	$= \frac{M_W}{G_W} = \frac{162}{1000} = 0.162\,\text{m}^3$

Volume of high range water reducer @ 0.8%M_C (V_{HR})	$= \dfrac{0.8 \times M_C}{100 \times G_{HR} \times 1000} = 0.00327\,\text{m}^3$
Volume of viscosity modifying agent @ 0.4%M_C(V_{VMA})	$= \dfrac{0.4 \times M_C}{100 \times G_{VMA} \times 1000} = 0.0017\,\text{m}^3$
Total volume of aggregates (V_{TA})	$= 1 - V_C - V_M - V_W - V_{HR} - V_{VMA} = 0.667\ \text{m}^3$
Volume of fine aggregates @37% of V_{TA} (V_{FA})	$= 0.37\ V_{TA} = 0.2504\ \text{m}^3$
Volume of coarse aggregates @63% of V_{TA} (V_{CA})	$= 0.63\ V_{TA} = 0.426\ \text{m}^3$

5. Determination of mass of materials for concrete mix

Mass of cement	= 441 kg
Mass of metakaolin to occupy the same volume	$= V_M \times G_M \times 1000$
	$= 0.0156 \times 2.6 \times 1000$
	= 40.57 kg
Mass of coarse aggregates	$= V_{CA} \times G_{CA} \times 1000 = 1191.19$ kg
Mass of fine aggregates	$= V_{FA} \times G_{FA} \times 1000 = 670.014$ kg
Mass of water	= 162 kg
Mass of high range water reducer	$= V_{HR} \times G_{HR} \times 1000 = 3.528$ kg
Mass of viscosity modifying agent	$= V_{VMA} \times G_{VMA} \times 1000 = 1.764$ kg

Case 2: Rubcrete

6. Determination of volume of materials in the rubcrete mix

   Here, the volume of fine aggregate to be replaced with crumb rubber is set as 15% (see Chap. 2). All other calculations remain the same as in the previous section.

Volume of fine aggregates @15% replacement	$= (1 - 0.15) \times V_{FA} = 0.213\ \text{m}^3$
Volume of crumb rubber (V_R)	$= 0.15 \times V_{FA} = 0.0375\ \text{m}^3$

7. Determination of mass of materials

Mass of fine aggregates	$= V_{FA} \times G_{FA} \times 1000 = 569$ kg
Mass of crumb rubber	$= V_R \times G_R \times 1000 = 24.42$ kg

Case 3: Rubcrete-Reinforced Using Polypropylene Fibres

Mix proportioning for polypropylene fibre reinforced rubcrete is presented with 15% crumb rubber and 0.2% polypropylene fibres as they were found to be optimum as discussed in Chap. 2.

Mass of polypropylene fibres: 0.002 × 0.91 × 1000 = 1.82 kg.

Other ingredients will now have to occupy 1 − 0.002 = 0.998 m^3 of the space.

8. Determination of volume of materials for polypropylene fibre-reinforced rubcrete

Volume of cement	$= \frac{M_C}{G_C \times 1000} \times 0.998 = 0.140\ m^3$
Volume of metakaolin	$= \frac{490 - 441}{3.14 \times 1000} \times 0.998 = 0.0156\ m^3$
Volume of water	$= \frac{M_W}{G_W} \times 0.998 = 0.1617\ m^3$
Volume of high range water reducer	$= \frac{0.8 \times M_C}{100 \times G_{HR} \times 1000} \times 0.998 = 0.00326\ m^3$
Volume of viscosity modifying agent	$= \frac{0.4 \times M_C}{100 \times G_{VMA} \times 1000} \times 0.998 = 0.00173\ m^3$
Total volume of aggregates	$= (1 - V_C - V_M - V_W - V_{HR} - V_{VMA})$ $\times 0.998 = 0.676\ m^3$
Volume of fine aggregates	= 0.3 × ×0.674 × 0.85 = 0.213 m^3
Volume of coarse aggregates	= 0.63 × 0.674 = 0.426 m^3

9. Determination of mass of materials for polypropylene fibre-reinforced rubcrete

Mass of cement	= 0.140 × 3.14 × 1000 = 440.11 kg
Mass of metakaolin	= 0.0156 × 2.6 × 1000 = 40.49 kg
Mass of fine aggregates	= 0.213 × 2.675 × 1000 = 568.92
Mass of coarse aggregates	= 0.426 × 2.79 × 1000 = 1189.91 kg
Mass of fine aggregates	= 0.213 × 2.675 × 1000 = 568.91 kg
Mass of water	= 0.1617 × 1000 = 161.7 kg
Mass of high range water reducer	= 0.00326 × 1.08 × 1000 = 3.52 kg
Mass of viscosity modifying agent	= 0.00173 × 1.02 × 1000 = 1.76 kg
Mass of rubber	= 0.37 × 0.676 × 0.15 × 0.65 × 1000 = 24.39 kg

Case 4: Rubcrete Reinforced Using Steel Fibres

Mix proportioning for steel fibre-reinforced rubcrete is presented with 15% crumb rubber and 0.75% steel fibres as they were found to be optimum as discussed in Chap. 2.

Mass of steel fibres: 0.0075 × 7.85 × 1000 = 58.875 kg.

Other ingredients will now have to occupy 1 − 0.0075 = 0.9925 m^3 of the space.

10. Determination of volume of materials for steel fibre-reinforced rubcrete

Volume of cement	$= \dfrac{M_C}{G_C \times 1000} \times 0.00925 = 0.139\ \text{m}^3$
Volume of metakaolin	$= \dfrac{490 - 441}{3.14 \times 1000} \times 0.00925 = 0.015\ \text{m}^3$
Volume of water	$= \dfrac{M_W}{G_W} \times 0.00925 = 0.161\ \text{m}^3$
Volume of high range water reducer	$= \dfrac{0.8 \times M_C}{100 \times G_{HR} \times 1000} \times 0.00925 = 0.003\ \text{m}^3$
Volume of viscosity modifying agent	$= \dfrac{0.4 \times M_C}{100 \times G_{VMA} \times 1000} \times 0.00925 = 0.002\ \text{m}^3$
Total volume of aggregates	$= (1 - V_C - V_M - V_W - V_{HR} - V_{VMA}) \times 0.00925 = 0.674\ \text{m}^3$
Volume of fine aggregates	= 0.37 × 0.674 × 0.85 = 0.212 m^3
Volume of coarse aggregates	= 0.63 × 0.674 = 0.425 m^3

11. Determination of mass of materials for steel fibre-reinforced rubcrete

Mass of cement	= 0.139 × 3.14 × 1000 = 437.69 kg
Mass of metakaolin	= 0.015 × 2.6 × 1000 = 40.269 kg
Mass of fine aggregates	= 0.212 × 2.675 × 1000 = 567.26 kg
Mass of coarse aggregates	= 0.425 × 2.79 × 1000 = 1187.31 kg
Mass of fine aggregates	= 0.212 × 2.675 × 1000 = 567.26 kg
Mass of water	= 0.161 × 1000 = 161 kg
Mass of high range water reducer	= 0.0032 × 1.08 × 1000 = 3.50 kg
Mass of viscosity modifying agent	= 0.0017 × 1.02 × 1000 = 1.75 kg
Mass of rubber	= 0.37 × 0.674 × 0.15 × 0.65 × 1000 = 24.32 kg

The same procedure can be followed for determination of mix proportions of fibre-reinforced rubcrete in mixes of different strength.

To summarise, for 1 m^3 ordinary concrete, rubcrete, rubcrete reinforced using polypropylene fibres and rubcrete reinforced using steel fibres discussed in this section, the mix proportions are:

Mass of materials for ordinary concrete	
Mass of cement	= 441 kg
Mass of metakaolin	= 40.57 kg
Mass of coarse aggregates	= 1191.19 kg
Mass of fine aggregates	= 670.014 kg
Mass of water	= 162 kg
Mass of high range water reducer	= 3.528 kg
Mass of viscosity modifying agent	= 1.764 kg
Mass of materials for rubcrete	
Mass of cement	= 441 kg
Mass of metakaolin	= 40.57 kg
Mass of coarse aggregates	= 1191.19 kg
Mass of fine aggregates	= 569 kg
Mass of crumb rubber	= 24.42 kg
Mass of water	= 162 kg
Mass of high range water reducer	= 3.528 kg
Mass of viscosity modifying agent	= 1.764 kg
Mass of materials for rubcrete reinforced using polypropylene fibres	
Mass of cement	= 440.11 kg
Mass of metakaolin	= 40.49 kg
Mass of fine aggregates	= 568.92 kg
Mass of coarse aggregates	= 1189.91 kg
Mass of fine aggregates	= 568.91 kg
Mass of water	= 161.7 kg
Mass of high range water reducer	= 3.52 kg
Mass of viscosity modifying agent	= 1.76 kg
Mass of rubber	= 24.39 kg
Mass of polypropylene fibres	= 1.82 kg
Mass of materials for rubcrete reinforced using steel fibres	
Mass of cement	= 437.69 kg
Mass of metakaolin	= 40.269 kg
Mass of fine aggregates	= 567.26 kg
Mass of coarse aggregates	= 1187.31 kg
Mass of fine aggregates	= 567.26 kg
Mass of water	= 161 kg
Mass of high range water reducer	= 3.50 kg
Mass of viscosity modifying agent	= 1.75 kg
Mass of rubber	= 24.32 kg
Mass of steel fibres	= 58.875 kg

References

1. Bureau of Indian Standards, Guidelines for concrete mix design proportioning. IS 10262: 2009, New Delhi, India (2009)
2. J.T. Ding, Z. Li, Effects of metakaolin and silica fume on properties of concrete. ACI Mater. J. **99**, 393–398 (2002). https://doi.org/10.14359/12222

A. Raj et al., *Rubcrete and Its Applications in Structures Subjected to Impact Loading*, SpringerBriefs in Applied Sciences and Technology,
https://doi.org/10.1007/978-981-95-6315-9

The manufacturer's authorised representative in the EU is Springer Nature Customer Service Centre GmbH, Europaplatz 3, 69115 Heidelberg, Germany. If you have any concerns regarding our products, please contact ProductSafety@springernature.com

Printed and bound by CPI Group (UK) Ltd, Croydon, CR0 4YY
07/07/2026
02160927-0004